Zafar Turakulov

The Dipole Radiation

Zafar Turakulov

The Dipole Radiation

Retarded Potentials and Maxwell Equations

LAP LAMBERT Academic Publishing

Impressum / Imprint

Bibliografische Information der Deutschen Nationalbibliothek: Die Deutsche Nationalbibliothek verzeichnet diese Publikation in der Deutschen Nationalbibliografie; detaillierte bibliografische Daten sind im Internet über http://dnb.d-nb.de abrufbar.

Alle in diesem Buch genannten Marken und Produktnamen unterliegen warenzeichen-, marken- oder patentrechtlichem Schutz bzw. sind Warenzeichen oder eingetragene Warenzeichen der jeweiligen Inhaber. Die Wiedergabe von Marken, Produktnamen, Gebrauchsnamen, Handelsnamen, Warenbezeichnungen u.s.w. in diesem Werk berechtigt auch ohne besondere Kennzeichnung nicht zu der Annahme, dass solche Namen im Sinne der Warenzeichen- und Markenschutzgesetzgebung als frei zu betrachten wären und daher von jedermann benutzt werden dürften.

Bibliographic information published by the Deutsche Nationalbibliothek: The Deutsche Nationalbibliothek lists this publication in the Deutsche Nationalbibliografie; detailed bibliographic data are available in the Internet at http://dnb.d-nb.de.

Any brand names and product names mentioned in this book are subject to trademark, brand or patent protection and are trademarks or registered trademarks of their respective holders. The use of brand names, product names, common names, trade names, product descriptions etc. even without a particular marking in this work is in no way to be construed to mean that such names may be regarded as unrestricted in respect of trademark and brand protection legislation and could thus be used by anyone.

Coverbild / Cover image: www.ingimage.com

Verlag / Publisher:
LAP LAMBERT Academic Publishing
ist ein Imprint der / is a trademark of
OmniScriptum GmbH & Co. KG
Bahnhofstraße 28, 66111 Saarbrücken, Deutschland / Germany
Email: info@omniscriptum.com

Herstellung: siehe letzte Seite /
Printed at: see last page
ISBN: 978-3-8443-0756-6

The Dipole Radiation.
Retarded Potentials and Maxwell Equations

Zafar Turakulov

ULUGH BEG ASTRONOMICAL INSTITUTE, UZBEK ACADEMY OF SCIENCES
E-mail address: tzafar@astrin.uz

ABSTRACT

A point-like dipole is the simplest and the most fundamental source of non-stationary electromagnetic field. The problem of field produced by this source and related problems are studied. The classical Hertz problem in which the dipole moment has fixed direction and oscillating magnitude, is actually equivalent to its variation, in which a dipole of constant magnitude rotates uniformly. Besides, there exists the problem of the field produced by an instantly polarized particle which is also equivalent the these two problems. The main feature of the latter is that its solution describes the phenomenon of free propagation of lines of force of fixed shape. All of them are studied first in the framework of the theory of retarded potentials, then by straightforward solving the Maxwell equations by the method of variables separation. The method is applied to other generalizations of the classical Hertz problem, which are problems of electromagnetic oscillations of a thin conducting rod and of magnetic burst of an axially-symmetric closed coil. The latter is in the scope because its solutions describe another case of free propagation of lines of force. Solutions obtained by this method are expressed in terms of Whittaker and Legendre functions and constitute a basis of "pure states" in the corresponding Hilbert spaces as in quantum mechanics.

Contents

To my mother.

Introduction

1. Point-like charge and point-like dipole in electrostatics

A point-like electric charge, suspended at rest, is the primary and the most fundamental source of electric field. The main law of electrostatic interaction has been formulated, first of all, for point-like charges. Interactions of complex charge distributions reduce to that of their parts which, after all, reduce to point-like charges. Thanks to the linearity of electromagnetic field, the potential of a point-like charge

$$(1.1) \qquad \Phi_q = \frac{q}{|\vec{r}|}$$

can be used as the Green function for the Poisson equation:

$$(1.2) \qquad \Phi(\vec{r}) = \int \frac{d\vec{r}'\mu(\vec{r}')}{|\vec{r} - \vec{r}'|}$$

where $\mu(\vec{r}')$ stands for spatial charge density, Φ is the potential produced by it, and the measure $d\vec{r}'$ in the integral is the element of volume of the space. Thanks to all this, the idea of point-like charge is reigning over electrostatics.

The second fundamental source of electrostatic field is a point-like dipole. This object was invented with use of passage to the limit of two opposite electric charges $\pm q$ placed in the endpoints of a vector $2\vec{a}$, when q tends to infinity and $|\vec{a}|$ does to zero so that the vector $\vec{d} = 2q\vec{a}$ called the dipole moment, has finite norm. A dipole with moment $\vec{d}$ produces the potential $\Phi_{\vec{d}}$ which is sum of potentials of its point-like charges

$$\Phi_{\vec{d}} = \Phi_q(\vec{r} + \vec{a}) - \Phi_q(\vec{r} - \vec{a}).$$

The potential of a point-like dipole can be obtained as the limit of this expression

$$\lim[\Phi_q(\vec{r} + \vec{a}) - \Phi_q(\vec{r} - \vec{a})] = 2\vec{d} \cdot \nabla \left(\frac{1}{|\vec{r}|}\right).$$

Hereafter we denote the operator

$$\vec{v} \cdot \nabla f \equiv \vec{v} \circ f$$

and call it the action of the vector $\vec{v}$ on the scalar f. Thus, by analogy to the potential of a point-like charge (1.1), a point-like dipole with moment $\vec{d}$ produces a potential equal to

$$(1.3) \qquad \Phi_{\vec{d}} = \vec{d} \circ \left(\frac{1}{|\vec{r}|} \right).$$

Though a point-like dipole is just a mathematical abstraction and does not exist in nature, it plays an important role in electrostatics. The laws of charge-dipole and dipole-dipole interactions are almost as fundamental as the Coulomb's law, besides, the law of interaction of two dipoles made it possible to understand interaction of two small magnets that was the first step towards creating magnetostatics.

2. Point-like charge and point-like dipole in electrodynamics

On a non-relativistic level, all electromagnetic perturbations propagate instantaneously, hence, potential of a dipole with varying dipole moment $\vec{d}(t)$ is

$$\Phi_{\vec{d}} = \vec{d}(t) \circ \left(\frac{1}{|\vec{r}|} \right).$$

No wave propagation process is seen in this expression. On a relativistic level such a dipole emits electromagnetic waves. To describe them, one has to solve Maxwell equations with $\vec{d}(t)$ as the source.

Adopting time as the fourth dimension unified not only space and time, but also electric and magnetic fields. In non-stationary state, these two fields cannot be decoupled from each other and obey Maxwell equations which they share alike. Fields of this kind contain electromagnetic waves unless they are produced by charges, magnets, and currents moving uniformly. The case of sources of electrostatic field moving uniformly is a subject of electrostatics, whereas the main subject of electrodynamics deals with emission, scattering and absorption of electromagnetic waves. Hereafter we consider only one of these three phenomena, namely, emission of electromagnetic waves.

On the classical level, electromagnetic waves have non-stationary sources. To study the phenomenon of radiation one needs the idea of the simplest and fundamental source of electromagnetic radiation, which would play the same role in this theory as point-like charge in electrostatics. The point-like charge itself is the first candidate for this role. Indeed, since everything consists of point-like charges, they are able to emit electromagnetic waves.

A point-like electric charge, suspended at rest and possessing varying magnitude would be the best candidate for the simplest and the most fundamental source of electromagnetic radiation. As a spherically-symmetric source it would emit spherically-symmetric waves. However, it is impossible because its magnitude is a conserved quantity. To emit radiation it must move non-uniformly and to describe its radiation one needs to find the field of arbitrarily moving charge. This problem seems to be quite solvable, because it is similar to the problem of magnetic field of a thin wire carrying direct current. The only difference is that the wire considered in the 3-space, is replaced by the time-like world curve of the charge. Indeed, like the vector of current, which is tangent of the wire everywhere and has constant norm equal to magnitude of the current, the 4-vector of current of a moving charge is everywhere tangent to its world line and has norm equal to the magnitude of the charge multiplied by c^2. So, before trying to solve the problem of electromagnetic field of an arbitrarily moving charge it would be reasonable to search for the solution of the spatial problem of magnetic field of thin wire in standard texts on classical electrodynamics. These texts provide only solutions for straight and circular conductors, since integrals can be done analytically only in exceptional cases, so, hardly one can hope to solve the same problem in space-time. Evidently, some simplification of the problem is needed.

The problem can be simplified by the following assumptions: the magnitude of the charge is very big and the amplitude of its oscillations about some fixed points is negligibly small. In fact, this is the same idea which led to the notion of point-like dipole. Indeed, passage to the limit $q \to \infty$, $\vec{a} \to 0$ where q is again magnitude of the charge and $\vec{a}$ stands for its deviation from the fixed point leads to the notion of point-like dipole with $\vec{d}(t) = q\vec{a}(t)$ as its time-depending dipole moment. As the result, we have a point-like source of electromagnetic field fixed in the space, which consists of infinitely big charge which can be ignored because it produces only a static field, and an oscillating dipole. It turns out that this time it is dipole, not charge whom the role of the simplest and the most fundamental source in electrodynamics, belongs.

3. The classical problem of Heinrich Hertz

A point-like dipole, suspended at rest, whose dipole moment depends on time, was introduced as the simplest and the most fundamental source of electromagnetic radiation by H. Hertz. He considered only the case of fixed direction of the dipole moment whose magnitude draws a harmonic oscillation with fixed frequency. Below, we call the problem of radiation from this kind of source "the classical Hertz problem". Besides, we are going to consider its modifications and call them "generalized Hertz problems". In this book we present the general solution of all of them.

The general solution of the classical Hertz problem provides solutions of all its generalizations due to linearity of the main equations. Indeed, if the spatial direction of the dipole moment is fixed, but its magnitude is specified as an arbitrary function of time, this function can be represented as a Fourier integral that reduces the problem to the classical one. If the dipole moment has no fixed direction in the space, it can be represented as a superposition of three dipoles with fixed directions and so is the field produced by them. This fact makes the Hertz dipole the simplest and the most fundamental source of electromagnetic radiation and the classical Hertz problem the most fundamental problem of the theory of electromagnetic radiation.

However, the Hertz dipole is not the only possible candidate for the most fundamental source of dipole radiation. There exist other fundamental sources. Let, for example, the magnitude of the dipole moment is fixed and the dipole rotates with constant angular velocity. Though this source is not very interesting from practical point of view, its magnetic analogue plays an important role in astrophysics as a model of a pulsar [1]-[3]. From mathematical point of view, this problem is as fundamental as the classical Hertz problem and all generalizations of the latter reduce to this as well as to that one.

And finally, there exists one more interesting source of dipole radiation, which also can be regarded as the fundamental one and which plays the central role in this book. It is a dipole whose dipole moment behaves as a step function of time. In practice it can be used as a model of a particle polarized or depolarized at some moment of time. Though all versions of the problem of dipole radiation considered above, are equivalent and reduce to each other, they require different approaches. Considerations of the field produced by a particle which was polarized at some moment of time, disclose some unexpected details of the phenomenon of dipole radiation and mathematical methods used in other cases.

4. Definition of retarded potentials

A point-like source of electromagnetic radiation, particularly, a dipole, suspended at rest, is an ideal object for employing the method of retarded potentials. By construction, retarded potential $A_i(x)$ is result of action of some linear operator on the source represented by its 4-current $J^i(x)$:

$$(4.1) \qquad A_i(x) = \int G_{ij}(x,y) \, J^j(y) \, \mathrm{d}^4 y.$$

The kernel of this integral operator $G_{ij}(x, y)$ is so-called Green function for the Maxwell equations. In spherical coordinates, one of Green functions for the d'Alembert equation has the form

$$(4.2) \qquad \delta(t^2 - r^2) = \frac{\delta(t - r)}{t + r} + \frac{\delta(t + r)}{t - r} \equiv \frac{1}{2}[G_{(ret)}(t, r) + G_{(adv)}(t, r)]$$

that is sum of retarded and advanced Green functions. Naturally, the retarded Green function acts in the future light cone $t = r$ and advanced Green function does so on the light cone of the past $-t = r$. They produce retarded and advanced potentials correspondingly. In case of point-like source, integration is trivial because the domain of integration is reduced to a single point, hence, so is the problem itself. However, theorists repeatedly encounter inexplicable difficulties when trying to solve it. First of all, to represent solutions of the Maxwell equations in the case of Hertz dipole, all authors but R.P. Feynman [4] (see, for example, the books [5]-[7]) divide the space into "near" or "static" and "wave" zones, though entire solution, if it is found, does not require the space to be divided any way. Description of the dipole radiation found in these books gives rise to some questions.

First, all authors write about electromagnetic fields, as if retarded Green function were unique. It is known, however, that there also exist Green functions of the form $1/(t - r)^2$ [9], which differ in the domain they are taken in. The three domains are: inside the light cone of the future, inside the light cone of the past or beyond the light cone, thus, even if the support of the Green function must belong to the light cone of the future, there is a two-parameter family of Green functions, whereas solution of the Maxwell equations vanishing at infinity is unique. So, there should be a rule to specify which combination is to be used. The second question is, what to do about a two-dimensional space of the source-related fields constructed by integrating Green functions and satisfying the same boundary conditions. In case of scalar field usage of the retarded Green function is undisputed, but this does not mean that in case of electromagnetic field it is exactly so for each component of electromagnetic field. The third question arises when analyzing explicit form of the source as current components when it turns out that none of authors noticed that representation of the source like $\vec{I} = \dot{\vec{d}}$ is incomplete (see the Chapt.1 for details).

Explicit form of the strengths obtained from retarded potentials for the Hertz dipole will be derived and analyzed below in the Sect. 2 of the Chapt. 1. The endpoints of the lines of force seen on the Fig.1, if they really exist, signify that there are electric charges in the space beyond the dipole, hence, due to the Maxwell equations, divergence of the electric strength there is non-zero (see the

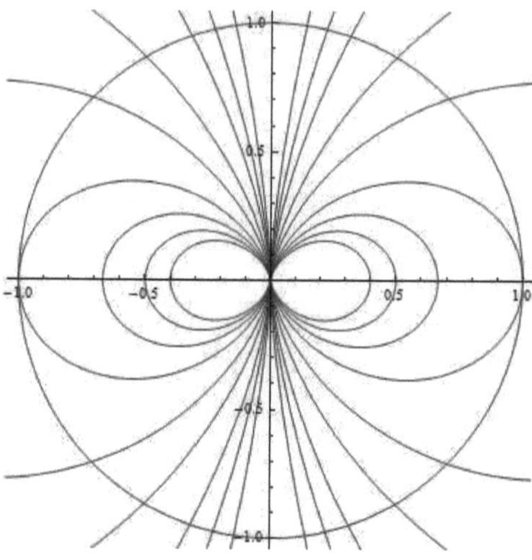

FIGURE 1. Lines of force of the dipole field and the sphere $r = t$.

Sect. 5 for details). It must be pointed out, however, that this picture was not obtained from analysis of the explicit form of retarded potential and exposes rather an intuitive pattern. Nevertheless, the fact that considerations which underlie the method of retarded potentials are able to lead to an expression for the electric strength $\vec{E}$ which does not satisfy at least one of Maxwell equations, namely, $\nabla \cdot \vec{E} = 0$, is somewhat harassing and require a detailed investigation.

5. Two distinct problems of the field of an instantly polarized particle

Almost any change of charge or current distribution in the space produces some electromagnetic radiation. The simplest change of this sort is instant polarization or depolarization of a particle. In this process, the particle dipole moment behaves as a step-function. In classical electrodynamics, dipole moment have usually been considered to behave as a smooth function of time, particularly, draw harmonic oscillations, as in case of Hertz dipole. In this section we consider the case when a particle, which has neither electric charge nor a dipole moment, is polarized at some moment of time $t = 0$ and produces the stationary dipole field under $t > 0$. Due to the causality principle, the field is non-zero only in the interior of the light cone $r \leq t$, hence, the field seems to be given by two well-known expressions, one for the interior of the light cone where the field is nothing but potential produced by a dipole, and one for its exterior, where the strengths are identically zero. The sphere $r = t$ expanding with speed of light seems to carry endpoints of lines of force, particularly, near the axis of symmetry. If it is so, the endpoints of lines of force of electric field signify presence of electric charge and the entire picture describes free propagation of these lines obeying the causality principle.

The notions of causality and lines of force seem to belong to different areas of physics: the earlier is known to be consequence of special relativity while the latter ordinarily characterizes stationary electric and magnetic fields. Non-stationary electromagnetic fields are usually being represented in the form of plane wave decompositions which provide no certain shape of lines of force. At the same time, there exist situations when a non-stationary field possesses certain lines of force. Below, we consider two situations of this sort in which behavior of lines of force is governed solely by causality. They occur when the field domain is restricted with causal boundary such that the field has certain stationary form inside some luminal surface and identically zero beyond it. This happens when some source, whose field in stationary case is known, and which is not a conserving value, is "switched on" or "off". If the lines of force break on the surface, one encounters the following paradox: on one hand, since lines of force have endpoints on the surface the latter carries corresponding electric, on the other hand, due to the problem under consideration, no luminal charge density should appear. Evidently, describing the phenomenon of free propagation of lines of force which are cut on some causal surface demands, first of all, to clear the matter up, whether it is so, and second, to resolve this paradox. It turns out that the problem is quite similar to the problem of shielding out a dipole by a sphere.

At each moment of time, the field is non-zero only inside a sphere about the particle, so, the source of the field looks as if it was shielded somehow by the surface. Evidently, to shield out the dipole, one needs the opposite dipole moment and to do it by the spherical surface, the opposite dipole should be distributed somehow on the sphere. As such, halves of the sphere must carry opposite charges that leads to a strange conclusion that there must be some non-zero electric charge distribution on the sphere $r = t$. On the other hand, the problem is defined quite clearly, to find out the filed only of a particle which is polarized at the moment of time $t = 0$. No other electric charge distribution in the space is assumed, therefore no other electric charges traveling with speed of light should appear, otherwise these considerations have nothing to do with solution of the initial problem. The difference is that in the original definition of the problem the total dipole moment in the space changes and due to the field specified by the retarded potential, it remains equal to zero and apparently provides solution of another problem, which could well be posed as follows. Let a particle decays into two parts, one is a polarized particle, another is a sphere expanding with speed of light and carries away the opposite dipole moment such that the sphere exactly shields out the new particle's dipole moment. Appearance of electric charge density in this case is expected and corresponds to the task definition. Thus, there exist, at least, two distinct problems of instant polarization of a particle, one simple (the total dipole moment changes) when a particle which had zero dipole moment becomes a dipole, and one complicated (the total dipole moment conserves) when the particle decays into two objects whose total moment is zero. Both of them are well-defined and admit correct solutions, hence, there exist two distinct solutions of the Maxwell equations, which provide solutions of these two problems. If the total dipole moment of the particle varies smoothly and the total moment in the space is always equal to zero, there must be some continuous charge density in the space to carry the moment away. It sounds somewhat strange, but in case of magnetic dipole this charge would be magnetic. It will be showed in this book that slight modification of retarded potentials, which could be called that of "retarded strengths" provides solution of the latter.

6. Maxwell equations, simple and double layers and gradient transformations

In this book we solve Maxwell equations

$$(6.1) \qquad \nabla \cdot \vec{E} = 4\pi\mu_e, \qquad \nabla \cdot \vec{H} = 4\pi\mu_m$$

$$\nabla \times \vec{H} = 4\pi\vec{I}_e + \frac{\partial\vec{E}}{\partial t} \qquad \nabla \times \vec{E} = -\frac{\partial\vec{H}}{\partial t} + 4\pi\vec{I}_m$$

where we have introduced densities of electric μ_e and magnetic μ_m charge and electric $\vec{I}_e$ and magnetic $\vec{I}_m$ currents. Though magnetic charge and its current do not exist, these terms are useful when analyzing expressions obtained when solving the Maxwell equations, particularly, obtained from retarded potentials. There exists a general rule in mathematics, which reads: whenever a differential equation is solved, the solution obtained must be checked by substituting it into the original equation. Doing so, we need to substitute expressions for electric and magnetic strengths obtained from retarded potentials into the Maxwell equations which contain all possible terms, to see, what electric and magnetic sources actually produce them.

In reality, magnetic lines of force have no endpoints, electric ones have only in electric charges. However, retarded potential of an instantly polarized particle provide strengths which are stationary inside the luminal sphere $r = t$, on which the lines of force apparently have endpoints (Fig.1). To explain this phenomenon we need to analyze stationary electric and magnetic fields possessing discontinuities. Discontinuities of stationary fields on a surface signify presence of their sources on it, which, in general, are surface densities of electric and magnetic charges and currents. The densities can be found from the Maxwell equations by using the theory of simple and double layers.

Unlike the ordinary electrostatic and magnetostatic problems, where boundary conditions can be specified by putting densities of magnetic charge and current equal to zero, in this case the filed is already known and it remains to extend this theory with these kinds of sources. The extended theory of simple and double sources yields the following equations

$$(6.2) \qquad \delta E_\perp = 4\pi\sigma_e, \quad |\delta\vec{E}_\parallel| = 4\pi|\vec{j}_m|, \quad (\delta\vec{E} \cdot \vec{j}_m)$$
$$\delta H_\perp = 4\pi\sigma_m, \quad |\vec{H}_\parallel| = 4\pi|\vec{j}_e|, \quad (\delta\vec{H} \cdot \vec{j}_e) = 0,$$

where $\delta\vec{E}$ and $\delta\vec{H}$ denote discontinuities of the strengths, σ_e and σ_m stand for surface densities of electric and magnetic charges, $\vec{j}_e$ and $\vec{j}_m$ l – surface densities of electric and magnetic currents on the surface.

Maxwell equations for electromagnetic waves in empty space have zero source terms which are (electric and magnetic) charge and current densities. The source-free Maxwell equations remain invariant under the duality transformation

$$(6.3) \qquad \vec{E} \to \vec{H}, \quad \vec{H} \to -\vec{E}.$$

If they are solved properly, substituting the strengths obtained into the equations (6.1) shows that the right-hand sides, which are electric and magnetic charge and current densities, are identically zero, otherwise they yield some non-zero expressions which are to be interpreted as the sources of the field needed to produce the strengths obtained this way. For example, if retarded potentials yield an expression for the electric strength $\vec{E}$ which has non-zero divergence, substituting it into one of the Maxwell equations yields certain density of electric charge in the space, equal to $\nabla \cdot \vec{E}$ divided by 4π. Correspondingly, the strengths obtained this way are produced by this charge density, hence, they are not strengths of electromagnetic wave in the empty space.

Explicit form of the electric strength obtained from the retarded potential for a Hertz dipole is found in the book [5], equation (9.18) on the page 271, in our notations:

$$\vec{H} = k^2(\vec{n} \times \vec{d})$$
$$\vec{E} = \vec{H} \times \vec{n}$$

where $\vec{n}$ is unit vector directed from the dipole. Note that the vector $\vec{H}$ is strictly azimuthal, thus has only φ-component, then the vector $\vec{E}$ possesses only θ-component. Now divergence of $\vec{E}$ can be calculated with use of the formula for divergence of a vector in spherical coordinates found in the book [**12**], equation (92) on the page 52. Divergence of this component has the form

$$\frac{1}{r \sin \theta} \frac{\partial}{\partial \theta}(E_\theta \sin \theta)$$

and since other components are zero, this expression is exactly $\nabla \cdot \vec{E}$. It can be zero only if E_θ is proportional to inverse $\sin \theta$. However, the expression for $\vec{E}$ contains only positive powers of $\sin \theta$. Consequently, $\nabla \cdot \vec{E} \neq 0$. Similarly, checking out the strengths obtained from retarded potential for magnetic dipole seem to yield a non-zero density of magnetic charge. This is another reason to include magnetic charge and current densities.

Usually, when considering electromagnetic waves, it suffices to introduce a purely space-like vector potential $\vec{A}$ whose derivatives provide both strengths in the form

$$(6.4) \qquad \frac{\partial \vec{A}}{\partial t} = \vec{E}, \quad \nabla \times \vec{A} = \vec{H}$$

Substituting this into the Maxwell equations (6.1) leads to something similar to the d'Alembert equation for the vector potential, but there is an important difference. The d'Alembert operator, as well as the Laplace operator, is defined for scalar functions whereas $\vec{A}$ is a vector, besides, double curl which appears this way, is not exactly the Laplace operator. Consequently, in general, Green functions for the d'Alembert equation cannot be used as if they are Green functions for the Maxwell equations and hence, solutions of the d'Alembert equation do not constitute a full-valued substitution for that of the Maxwell equations. That is why it happens that retarded potentials provide wrong solutions of the source-free Maxwell equations.

As seen from the equations (6.4), the vector potential is ambiguous. So-called gradient transformation

$$(6.5) \qquad \vec{A} \to \vec{A} + \nabla \Phi$$

where Φ is an arbitrary scalar function, leaves both the strength invariant. This transformation will be used when constructing the field of an instantly polarized particle.

7. Brief contents of this book

Since in this book we emphasize the mathematical aspects of the subject, it is convenient to use dimensionless system of units. So, we put the speed of light equal to one and, suppose that all lengths are related to some unit of length, so that all distances are dimensionless. When considering electromagnetic radiation possessing a certain frequency ω we use it as the inverse unit of length. This allows us to write expressions like $\delta(t - r)$ despite that the argument of such a distribution must be strictly numerical.

In the Chapt. 1, the exact retarded potential for a Hertz dipole with given magnitude, direction and frequency is constructed and it is shown that the strengths obtained from it do not satisfy the source-free Maxwell equations. Then the phenomenon of instant polarization of a point particle is studied. As the result, the theory of retarded potentials is modified and it turned out that modified theory provides solution of the problem of instant polarization under which the opposite dipole moment runs away with speed of light. In the Chapt. 2, we consider an auxiliary problem of complete shielding of an arbitrary source of stationary electric and magnetic field by surface densities of electric charge and current on a given surface. Afterwards, we use these results to introduce another modification of the notions of retarded potential and causal shielding of an instantly polarized particle. The result actually provides solution of the problem of the field of an instantaneously polarized particle. In the Chapt. 3, the field of a Hertz dipole is considered. Analytical solutions of the Maxwell equations are found which show that the method of retarded potentials provides only an asymptotical solution. The field of the field of rotating dipole is found and it is shown that the total angular momentum of the field is zero. Then we pass to the problem of electromagnetic oscillations of a conducting rod. Solutions of the Maxwell equations for this case are obtained in prolate spheroidal coordinates. In the Chapt. 4, we outline a solution of the Maxwell equations free propagating of closed lines of force. Each of these chapters contains a special section entitled "Conclusions", in which the main results are briefly formulated. The main results of our research, presented in this book, are outlined in the Chapt 5. which is also entitled "Conclusions", but in less formal manner than in the closing sections of the previous chapters. Besides, the book contains a special chapter entitled "Appendices" in which we demonstrate the main mathematical methods used for reducing the Maxwell equations to ordinary differential equations.

Acknowledgements

The author is grateful to J. Kijowski, for organizing a seminar in IFPAN, Warsaw in 1999, where results presented in this book have been presented for the first time, to participants of this seminar J. Mostowski and A. Orlowski for interesting discussions and U. Mutze for valuable remarks on the material of this book.

CHAPTER 1

The retarded potential of a dipole and the method of retarded strengths

1. Hertz dipole as a source of an electromagnetic field

In Maxwell equations (6.1), the source of electromagnetic field is represented by certain densities of electric and magnetic charges and currents, whereas a Hertz dipole provides only its dipole moment $\vec{d}(t)$. Therefore, to solve these equations, we need, first of all, to represent this source in the standard form of charge and current densities, which corresponds to this dipole moment. Usually, one identifies this source with its current $\vec{I} = \dot{\vec{d}}$ [4]-[7] which possesses single spatial component, say, $I_z = \omega d_z \delta^3(r) \sin \omega t$ where d_z is magnitude of the dipole moment, ω is its frequency. It must be pointed out, however, that the dipole possesses also a non-zero charge density because any straight current confined in a restricted segment inevitably accumulates electric charge in its endpoints. Consequently, a Hertz dipole possesses a non-zero oscillating charge density which must be taken into account as well as the current. In this section we find out this density for a given Hertz dipole.

To find this density, note that the current $\vec{I}$ and charge μ densities satisfy the following conservation law:

$$(1.1) \qquad \frac{d\mu}{dt} + \nabla \cdot \vec{I} = 0$$

from which μ can be found as follows. First, note that the 3-dimensional δ-function takes different form in different coordinate systems. For example, in round cylinder coordinates $\{z, \rho, \varphi\}$ with the measure of volume integration $dz \cdot d\rho \cdot \rho \, d\varphi$ the 3-dimensional δ-function is $(\pi\rho)^{-1}\delta(z)\delta(\rho)$ because the domain of integration over ρ starts at $\rho = 0$ that yields the factor $1/2$. In case of Hertz dipole with fixed spatial direction this coordinate system is the most convenient because the current has single spatial component and so is the vector potential (6.4). Now, divergence of $\vec{I}$ is

$$(1.2) \qquad \nabla \cdot \vec{I} = \frac{\omega d_z}{\pi\rho}\delta'(z)\delta(\rho) \sin \omega t,$$

21

consequently, the charge density is

$$
(1.3) \qquad \mu = \frac{d_z}{\pi \rho} \delta'(z) \delta(\rho) \cos \omega t.
$$

Derivative of the δ-function seems to be somewhat unusual kind of charge density. Nevertheless, it is exactly in accord with the passage to the limit which underlies the notion of a point-like dipole considered in the previous chapter. To show this, let us consider electrostatic potential produced by this charge density as it appears in the equation (1.2). For simplicity, we put the dipole moment equal to one, thus, the potential has the form

$$
\Phi = \int\limits_{-\infty}^{\infty} dz' \int\limits_{0}^{\infty} \rho' \, d\rho' \int\limits_{0}^{2\pi} d\varphi \frac{\delta'(z')\delta(\rho')}{\pi \rho' \sqrt{(z-z')^2 + \rho^2}} =
$$

$$
= \int\limits_{-\infty}^{\infty} dz' \frac{\delta'(z')}{\sqrt{(z-z')^2 + \rho^2}} = \frac{\partial}{\partial z} \frac{1}{\sqrt{z^2 + \rho^2}}
$$

that is exactly coincides with the equation (1.3) with unit dipole moment. Thus, finally, we have obtained both charge and current densities for a Hertz dipole, which are

$$
(1.4) \qquad I_z = \frac{\omega d_z}{\pi \rho} \delta(z)\delta(\rho) \sin \omega t, \quad \mu = \frac{d_z}{\pi \rho} \delta'(z)\delta(\rho) \cos \omega t.
$$

Now, let us consider a point-like magnetic dipole. Such a dipole can be obtained as the limit of an infinitesimally-thin conducting ring lying in the plane $z = 0$, carrying electric current I. Magnetic moment of such a ring is equal to aI, where a is its radius, so, to get a point-like source we need to pass to the limit $a \to 0$, $I \to \infty$ such that their product is finite. The main feature of this limit is that the current density cannot be concentrated in the point because afterwards we multiply it by the distance from the center. In this situation the current density j can only be defined as derivative of the δ-function,

$$
(1.5) \qquad j = \frac{\delta'(\rho)}{\rho}
$$

because only in this case we have the proper form of the density of magnetic moment which after integration yields the value of magnetic moment (equal to one in this case).

Now, let us look, what potential is produced by this current density. Multiplying the density by the same Green function and integrating yields:

$$
(1.6) \qquad |\vec{A}| = \int\limits_{-\infty}^{\infty} dz' \int\limits_{0}^{\infty} \rho'\, d\rho' \int\limits_{0}^{2\pi} d\varphi \frac{\delta(z')\delta'(\rho')}{\pi\rho'\sqrt{(z-z')^2+\rho^2}} =
$$

$$
= \int\limits_{-\infty}^{\infty} dz' \frac{\delta'(z')}{\sqrt{(z-z')^2+\rho^2}} = \frac{\partial}{\partial\rho}\frac{1}{\sqrt{z^2+\rho^2}}.
$$

It must be pointed out that in this case we misuse the method of Green functions. The point is that we apply the Green function for a scalar field to a vector. Usually, this procedure is being justified by the arguments found in the M. Born and E. Wolf monograph [8] which consists in the following. In standard Cartesian coordinates, the operation $\nabla \times \nabla \times$ (double curl) can be replaced by the Laplacian due to the identity

$$
\nabla \times \nabla \times \vec{A} = \nabla(\nabla \cdot \vec{A}) - \triangle\vec{A},
$$

proviso that the vector has zero divergence. The same can be done in cylindric coordinates only if the operations are applied to the component A_z of the vector, because this component is, actually, a Cartesian one. In case case of φ-component whose divergence is zero, its minus double curl is not equal to its ordinary Laplacian. Nevertheless, the result obtained coincides with the vector potential of a point-like magnetic dipole given by the equation (3.1) of the next chapter. This coincidence will be discussed in the Chapt.3. So, it looks like that retarded potential can be used in the case of an instantly magnetized particle.

2. The retarded potential of a Hertz dipole

To obtain the explicit form of the retarded potential for a Hertz dipole of unit magnitude and frequency ω with direction fixed in the space as the z-axis, we substitute the densities given by the equations (1.4) into the equations (4.1) and (4.2) and integrate them. In round cylinder coordinates the retarded potential possesses two non-zero components which are the time component Φ and the z-component A_z produced by charge and current densities correspondingly. In both cases the integrand contains either product of two δ-functions or of one δ-function and one its derivative, which depend on different variables. This fact creates no problems because the domain of integration is 4-dimensional and integration over the time coordinate removes one of them. In both cases we use the retarded Green function in the form

$$G = \frac{2\delta\left(t - t' - \sqrt{(z - z')^2 + \rho^2}\right)}{(t - t') + \sqrt{(z - z')^2 + \rho^2}}$$

which is valid only for singular sources concentrated on the axis $\rho = 0$, hence, the variable of integration ρ' is identically zero and we do not need to include it into the Green function. For simplicity, we put again the magnitude of the dipole equal to one.

To obtain explicit form of the time component of the potential Φ, consider the integral over 4-dimensional domain

$$\Phi = \int\limits_{-\infty}^{\infty} dt' \int\limits_{-\infty}^{\infty} dz' \int\limits_{0}^{\infty} \rho'\,d\rho' \int\limits_{0}^{2\pi} d\varphi' \times$$

$$\times \quad \frac{\delta'(z')\delta(\rho')}{\pi\rho'} \cos\omega t' \, \frac{2\delta\left(t - t' - \sqrt{(z - z')^2 + \rho^2}\right)}{(t - t') + \sqrt{(z - z')^2 + \rho^2}}.$$

After integrating over the time and radial coordinates where the value of t' fixes as

$$(2.1) \qquad\qquad t' = t - \sqrt{(z - z')^2 + \rho^2}$$

(retarded time). Now, integration over z actually differentiates the integrand on this variable, and the φ integration is trivial, and as the result, we have

$$\Phi = \frac{\partial}{\partial z} \frac{\cos\omega\left(t - \sqrt{z^2 + \rho^2}\right)}{\sqrt{z^2 + \rho^2}}.$$

Finally,

$$\Phi = \frac{\omega z \sin\omega(t - r)}{r^2} - \frac{z \cos\omega(t - r)}{r^3}$$

where r stands for the distance from the dipole, which is equal to $\sqrt{z^2 + \rho^2}$. Surprisingly enough, that the field which was believed to be purely radiative, possesses a non-zero time-like component of the vector potential. The next task is to obtain the component A_z. For this end, we consider another integral over

4-dimensional domain,

$$A_z = \int\limits_{-\infty}^{\infty} \mathrm{d}t' \int\limits_{-\infty}^{\infty} \mathrm{d}z' \int\limits_{0}^{\infty} \rho'\,\mathrm{d}\rho' \int\limits_{0}^{2\pi} \mathrm{d}\varphi' \times$$

$$\times \;\; \frac{\omega}{\pi\rho'}\delta(z')\delta(\rho')\sin\omega t'\; \frac{2\delta\left(t - t' - c^{-1}\sqrt{(z - z')^2 + \rho^2}\right)}{(t - t') + \sqrt{(z - z')^2 + \rho^2}}.$$

In this case integration is simpler because the integrand does not contain derivative of the δ-function, thus, the result does not contain the differentiation operator. Again, integration over time and radial coordinates is trivial and replaces t' with the expression (2.1) and finally, we obtain the well-known expression for the desired component

$$A_z = \frac{\omega\sin\omega(t - r)}{r}.$$

As the result, we have the complete retarded potential of Hertz dipole of unit magnitude in the z-direction and frequency ω:

$$(2.2) \qquad \Phi = \frac{\omega z\sin\omega(t - r)}{r^2} - \frac{z\cos\omega(t - r)}{r^3}$$

$$A_z = \frac{\omega\sin\omega(t - r)}{r}.$$

Appearance of the time component Φ of the vector potential shows that representation of the electric strength in the equation (6.4) is incomplete. Moreover, ignoring it leads to an apparently wrong expression of the field of dipole under $\omega = 0$. Contrary to expectations, the potential obtained does not look like that of electromagnetic wave. Its time component contains two terms, one proportional to ω/r and one proportional to $1/r^2$, none of which possesses properties of radiation, though the z-component of the vector potential coincides with that of non-relativistic representation of the potential of the Hertz dipole [4]-[7].

The strengths of the electromagnetic field which correspond to the retarded potential can now be obtained, but, in fact, they are of no use because they do not satisfy the source-free Maxwell equations under $r \neq 0$. To see this it suffices to consider one of them, namely, the source-free Maxwell equation $\nabla \cdot \vec{E} = 0$ that contains Laplacian of Φ and the second t and z derivative of A_z. Evidently, the left-hand side of this equation contains four different functions multiplied by powers 0 to 3 of ω, which must disappear separately because this equation is valid under arbitrary ω. Nevertheless, two of them come out identically zero and two others have amplitudes linear on ω and ω^2. It turns out that both of them are non-zero and electric charge density needed to produce the retarded potential (2.2) is

$$(2.3) \qquad \mu = \frac{2\omega^2 z}{4\pi r^3}\,\omega^2 \sin\omega(t-r) - \frac{2\omega z}{4\pi r^4}\sin\omega(t-r)$$

(.4). Consequently, electric strength obtained from the retarded potential for a Hertz dipole, does not satisfy at least one of the source-free Maxwell equations and hence, this expression has nothing to do with radiation from this object. The non-zero divergence of the electric strength exposes presence of electric charge density in the space which can be calculated from the corresponding Maxwell equation (6.1). Note, that the duality transformation (6.3) replaces the electric dipole with magnetic one and, correspondingly, specifies a non-zero magnetic charge density in the space which can only produce magnetic strength obtained from the retarded potential.

3. Retarded potential of an instantly polarized particle

Consider a point particle which has zero electric charge and magnetic and dipole moments, whose state changes at the moment of time $t = 0$ such a way that its dipole moment becomes non-zero, thus, behaves like the well-known step function $\vartheta(t)$. This phenomenon is called polarization of the particle. Below we consider electromagnetic field of a particle polarized at the moment of time and assume that after that its dipole moment $\vec{d}$ remains constant in its magnitude and direction. To obtain the field in question, we need to solve the Maxwell equations with the dipole moment as the source, thus, first of all, to specify the source as electric charge and current densities. The charge density of a point dipole has the form (1.3), hence, the charge density of the particle under consideration has the form

$$\mu = \frac{d_z}{\pi\rho}\delta'(z)\delta(\rho)\vartheta(t)$$

and it remains to find out the corresponding current density from the equations (1.1,1.1). Substituting this charge density into the equation leads to the following solution:

$$(3.1) \qquad \mu = \frac{d_z}{\pi\rho}\delta'(z)\delta(\rho)\vartheta(t), \quad I_z = -\frac{d_z}{c\pi\rho}\delta(z)\delta(\rho)\delta(t).$$

The time and z-components of the retarded potential for an instantly polarized particle have the form of integrals considered above, thus,

$$\Phi = \int\limits_{-\infty}^{\infty} dt' \int\limits_{-\infty}^{\infty} dz' \int\limits_{0}^{\infty} \rho'\, d\rho' \int\limits_{0}^{2\pi} d\varphi' \times$$

$$\times \; \frac{d_z}{\pi\rho'}\delta'(z')\delta(\rho')\vartheta(t')\, \frac{2\delta\left(t - t' - c^{-1}\sqrt{(z - z')^2 + \rho^2}\right)}{(t - t') + \sqrt{(z - z')^2 + \rho^2}}.$$

Again, integration over the variable t' yields the equality (2.1) with the additional restriction $ct \geq r$ so that

$$\Phi = \int\limits_{-\infty}^{\infty} dz'\delta'(z')\frac{d_z}{r} = d_z\frac{\partial}{\partial z}\left(\frac{\vartheta(t - r)}{r}\right)$$

that is almost exactly the dipole potential "switched on" at the moment of time $t = 0$. The difference is that the potential contains an extra δ-function singularity which appears when differentiating the ϑ-function. Hereafter we ignore this singularity considering the retarded potential everywhere beyond the two sources of the electromagnetic field in question, which are confined only in the origin of coordinates and the sphere, thus, have measure zero.

In any case, the lines of force are cut off by the sphere $r = t$. This representation is incomplete, because there also is the non-zero current density I_z given by one of the equations (3.1), which produces the corresponding z-component of the vector potential. To obtain it, we need to calculate the following integral:

$$A_z = -\int\limits_{-\infty}^{\infty} dt' \int\limits_{-\infty}^{\infty} dz' \int\limits_{0}^{\infty} \rho'\, d\rho' \int\limits_{0}^{2\pi} d\varphi' \times$$

$$\times \; \frac{d_z}{c\pi\rho'}\delta(z')\delta(\rho')\delta(t')\, \frac{2\delta\left(t - t' - c^{-1}\sqrt{(z - z')^2 + \rho^2}\right)}{(t - t') + \sqrt{(z - z')^2 + \rho^2}}.$$

In this case integration is trivial and finally we have the following representation of the retarded potential of a dipole "switched on" at $t = 0$:

$$\Phi = d_z\frac{\partial}{\partial z}\left(\frac{\vartheta(t - r)}{r}\right)$$

$$A_z = \delta(t - r)\frac{d_z}{r}.$$

It is seen that this form of potentials produces too many singularities. Applying the d'Alembert operator to its time component Φ yields the single non-zero term equal to $\delta(t-r)$ multiplied by the radial component of $\nabla\Phi$, that is exactly 4π multiplied by the surface charge density which corresponds to the endpoints of the lines of force. The term proportional to $\delta'(t-r)$ is identically zero because so is combination of the second derivatives in the d'Alembertian of any function of single variable $(t-r)$. Hence, in this case the retarded potential does not provide any solution of the source-free Maxwell equation and it remains unclear, whether it yields solution of the problem of a particle decay into a polarized particle and anti-polarized expanding sphere $r = t$. However, such a decay can be described by retarded strengths which can be obtained by replacing the retarded potentials with the modified expressions

$$(3.2) \qquad \Phi = d_z \vartheta(t-r) \frac{\partial}{\partial z}\left(\frac{1}{r}\right)$$

$$A_z = \delta(t-r)\frac{d_z}{r}$$

and defining the retarded strengths as follows:

$$(3.3) \qquad \vec{E}_{ret} = d_z \vartheta(t-r)\,\nabla\frac{\partial}{\partial z}\left(\frac{1}{r}\right) - \delta(t-r)\frac{\partial}{\partial t}\frac{d_z}{r}\vec{n}_z$$

$$(\vec{H}_{ret})_\varphi = \delta(t-r)\frac{\partial}{\partial\rho}\frac{d_z}{r}$$

(here the φ-component of the magnetic strength is equal to its length, n_z is the unit vector collinear with the z-axis).

4. Retarded strengths of an instantly polarized particle

As expected, electromagnetic field of an instantaneously polarized particle is stationary inside the sphere $r = t$ and zero beyond it, thus, it has certain lines of force inside the sphere and the entire solution should describe the process of free propagation of the lines of force. However, at the moment, it is not clear, whether retarded potentials describe this phenomenon or something else, namely, the dipole moment carried away by the sphere expanding with speed of light. If there are no sources of electromagnetic field under $r \neq 0$, retarded potentials describe the earlier, otherwise it does something else. In this section we analyze the strengths obtained from the retarded potential and show that they do not satisfy the source-free Maxwell equations. Instead, they expose presence a source of the field on the expanding sphere.

Inside the sphere, the electric field is exactly as the field of a stationary dipole and on the sphere its lines of force are cut, thus have endpoints and it is easy to calculate the corresponding surface charge density from one of equations (6.2). Due to these equations, finite jumps of the normal and tangential components of the strengths signify presence of some sources of the fields, which reveal as surface densities of charges and currents. In spherical coordinates $\{r, \theta, \varphi\}$ the normal component of the electric strength inside the sphere $r = t$ is

$$E_\perp = \frac{\partial \Phi}{\partial r} = \frac{d_z \cos \theta}{r^3}$$

and since beyond the sphere the strength is zero, this expression specifies the jump of $E_\perp$, hence, the surface density of electric charge on the sphere has the form

$$\sigma_e = \frac{d_z \cos \theta}{4\pi r^3}.$$

Note that integral of this electric charge distribution multiplied by $r \cos \theta$, over the sphere yields the total dipole moment on the sphere. Since this quantity does not depend on r, it remains constant on the expanding sphere that signifies that the charge density carries away the dipole moment of the polarized particle. Besides, there is also jump of the tangential component of the electric strength which on the sphere is equal to

$$|\vec{E}_\|| = \frac{1}{r}\frac{\partial \Phi}{\partial \theta} = -\frac{d_z \sin \theta}{r^2}.$$

Since on the outer side of the sphere the electric strength is zero, this expression specifies the jump of the tangential component. Substituting it into the corresponding equation (6.2) yields the surface density of magnetic current orthogonal $\vec{E}_\|$. Alternatively, we can consider the entire form of the Maxwell equations and check out if this jump can be explained by the δ-function singularity of toroidal magnetic strength (3.3). The tangential component of the magnetic strength has the form of $\delta(t - r)$ multiplied by the function

$$\frac{\partial A_z}{\partial \rho} = -\frac{d_z \rho}{r^3} = -\frac{d_z \sin \theta}{r^2}$$

that is exactly the jump of $\vec{E}_\|$. In fact, expansion of the sphere with speed of light replaces the magnetic current with the corresponding current of displacement. This phenomenon will be used in the next chapter.

5. Conclusion

In this chapter, we tried to apply the method of retarded potentials to the problem of electromagnetic field of an instantaneously polarized particle. Since, due to the causality principle, the lines of force are cut on the sphere $r = t$, which expands with speed of light, we presumed that the solution to be found describes the phenomenon of decay of an originally non-polarized particle into a polarized particle and an object which is an infinitesimally-thin sphere expanding with speed of light and which carries away the opposite dipole moment such that the total dipole moment in the space remains equal to zero. The explicit form of the retarded potential is found and it turns out that it produces too many singularities when calculating the field strengths. Therefore we decided to modify the theory of retarded potentials replacing it with the theory we called "the theory of retarded strengths". The theory can be formulated completely by defining retarded strengths obtained from the vector potential, by differentiating only the Green function factor, but such a theory is actually of no use. Analysis of the explicit form of retarded strengths shows that they satisfy Maxwell equations for electromagnetic field produced by two sources, one of which is an instantly polarized particles, another – the infinitesimally-thin sphere which carries away the opposite magnetic moment and expanding with speed of light. Unlike retarded strengths, retarded potential creates extra singularities. Indeed, differentiating its components (3.2) yields strengths whose components contain derivatives of the δ-function. As for the retarded strengths, they describe the field of this two objects and, thereby, solution of the problem of the field of decay of an initially non-polarized particle into the two objects.

Now we can resume our conclusions.

(1) Components of charge and current densities of a Hertz dipole provided in almost all books on classical electrodynamics $(\vec{I} = \dot{\vec{d}})$ is wrong. The right representation which contains also the charge density, is presented by the equations (1.4).

(2) Retarded potential of a Hertz dipole provided in almost all books on classical electrodynamics which does not contain the time component Φ of the potential, is wrong. The right representation which contains both spatial and time components of the potential, is presented by the equations (2.2).

(3) Neither the complete retarded potential, nor its generally accepted incomplete form provide solutions of the source-free Maxwell equations. At least one of these equations, namely, $\nabla \cdot \vec{E} = 0$, is broken. Electric strength as it is found from retarded potential, corresponds to electric

charge density presented in the equation (2.3). Consequently, retarded potentials have nothing to do with solutions of the Maxwell equations and need, at least to be modified.

Now, we pass to the original problem of the field of an instantly polarized particle without luminal charges. It turns out that solution of this problem requires another modification of the theory of retarded potential, but, before doing it, something else is to be done. The point is that in this case the problem of endpoints of the lines of force is to be solved without appearance of any extra sources of the electromagnetic field. In other words, there must not be endpoints of the lines and the problem should be solved in terms of source-free solutions of the Maxwell equations. However, to do it, we need first to study the terms to be eliminated. Therefore, first, we consider the auxiliary problem of screening of an arbitrary stationary source of electric and magnetic field.

CHAPTER 2

The field of instantly polarized particle

1. The problem definition

In this chapter we expose our attempts to modify the method of retarded potential such that it gives correct solutions of the Maxwell equations for an instantly polarized particle. Below, we show how to eliminate electric charge and current densities on the sphere $r = t$ in this case. Anticipating things, we must warn that, though such a modification exists, it does not give the desired solutions because there are other Maxwell equations ignored before, which are also to be solved. In fact, this chapter exposes a failed attempt to make the method of retarded potentials work. Correct solutions of the Maxwell equations are obtained different way in the next chapter, so, those who seeks for these solutions, can omit this chapter without lose of practically useful information.

To eliminate electric charge and current densities on the sphere $r = t$, we return electro- and magnetostatics and consider the following problem. Let there be stationary densities of electric charge and current confined wholly inside a spatial domain G and let there be a closed surface Σ surrounding this domain. The task is to place such surface densities of charge and current on this surface that the total field produced by charges and currents in the domain G and on the surface Σ is identically zero everywhere outside this surface. This phenomenon is called shielding of the sources confined in the domain G. After this problem is solved, we pass to the phenomenon of causal shielding which takes place in the phenomenon of the field of instantly polarized particle.

Now, let us formulate this problem mathematically. For this end, we introduce two domains in the space, G_1 and G_2, such that G_1 is a part of G_2, assuming that the sources are confined in G_1 and Σ is the boundary of the domain G_2, or, using notations of the set theory,

$$G_1 \subset G_2, \quad \Sigma = \partial G_2.$$

Thus, the sources are confined in the domain G_1, and on the surface ∂G_2, the fields produced by these sources are non-zero in G_2 and identically zero for every point P beyond G_2:

$$P \notin G_2 \Rightarrow \vec{E}(P) \equiv 0, \quad \vec{H}(P) \equiv 0.$$

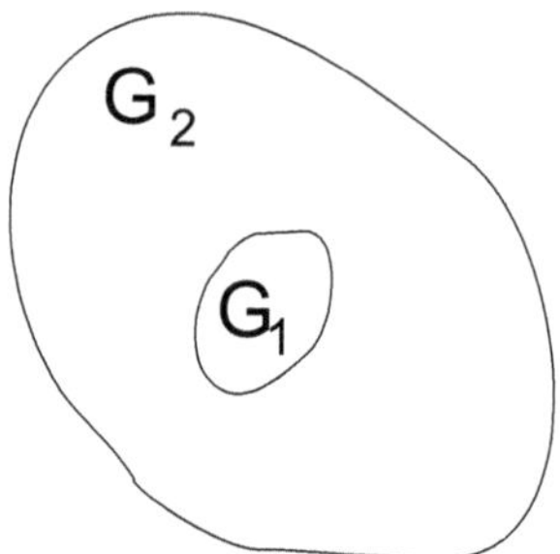

FIGURE 1. Definition of the problem of shielding.

Then the lines of force of the electrostatic field $\vec{E}$ start in G_1 and end up in G_2, exposing some surface charge density on ∂G_2, whereas lines of force of the magnetostatic field $\vec{H}$ touch this surface without crossing it. In fact, this condition signify that the field $\vec{H}$ is tangent to the surface ∂G_2 everywhere. At the same time both of these fields must satisfy the stationary source-free Maxwell equation in the layer $G_2 \backslash G_1$

$$(1.1) \qquad \nabla \cdot \vec{E} = \nabla \cdot \vec{H} = 0, \quad \nabla \times \vec{E} = \nabla \times \vec{H} = 0.$$

It is convenient to distinguish the fields produced by the original sources, which are confined inside the domain G_1 and the fields produced by the shielding sources which are distributed on the surface ∂G_2. Hereafter $\vec{E}_1$ and $\vec{H}_1$ stand for electric and magnetic fields produced by the main sources, and $\vec{E}_2$ and $\vec{H}_2$ do for the fields produced by the shielding ones. Thus, the sources of the fields $\vec{E}_1$ and $\vec{H}_1$ are confined in in the domain G_1 and sources of the fields $\vec{E}_2$ and $\vec{H}_2$ lie wholly on the surface ∂G_2. By construction, all of them satisfy the equations (1.1) in the layer between the surfaces ∂G_1 and ∂G_2 (Fig.1) and, by definition, their sums $\vec{E}_1 + \vec{E}_2$ and $\vec{H}_1 + \vec{H}_2$ are zero in the outer space. All the strengths $\vec{E}_1$, $\vec{E}_2$, $\vec{H}_1$ and $\vec{H}_2$ satisfy the equations (1.1) everywhere beyond the domain G_1, but the strengths $\vec{E}_2$ and $\vec{H}_2$ have finite discontinuity on the surface ∂G_2. These discontinuities are specified by the surface densities in accord with the equations (6.2), in which we need to put the surface densities of magnetic charge σ_m and magnetic current $\vec{j}_m$ equal to zero for physical reasons. Thus, the discontinuities are given by the equations

$$(1.2) \qquad \delta E_{2\perp} = 4\pi\sigma_e, \qquad |\delta\vec{E}_{2\|}| = 0$$

$$(1.3) \qquad \delta H_{2\perp} = 0, \qquad |\vec{H}_{2\|}| = 4\pi|\vec{j}_e|, \quad (\delta\vec{H}_2 \cdot \vec{j}_e) = 0,$$

so, hereafter we omit the subscripts e. These equations specify the boundary conditions of the problem.

2. Electrostatic and magnetostatic potentials

Since, due to the equations (1.1), both electrostatic and magnetostatic fields are potential ones, they possess electric and magnetic potentials Φ and Ψ in the layer $G_2 \backslash G_1$ such that

$$\vec{E}_2 = \nabla\Phi, \quad \vec{H}_2 = \nabla\Psi,$$

which satisfy Laplace equation:

(2.1) $$\triangle\Phi = \triangle\Psi = 0.$$

with boundary conditions conditions discussed below.

Replacing the strengths produced by the shielding sources with their representation in terms of potentials in the equation (1.2) yields the following boundary conditions for the potentials:

(2.2) $$(\nabla\Phi)_{\|}|_{\partial G_2} = -\vec{E}_{1\|}, \quad (\nabla\Psi)_{\perp}|_{\partial G_2} = -H_{1\perp}.$$

These boundary conditions combined with the condition that both potentials vanish at infinity, make it possible to find certain solutions of the Laplace equations for the potentials (2.1) that signifies existence and uniqueness of the potentials Φ and Ψ. After these potentials are found, all the rest components of the fields $\vec{E}_2$ and $\vec{H}_2$ on the surface can be obtained as

$$\vec{E}_{2\perp} = (\nabla\Phi)_{\perp}|_{\partial G_2}, \quad H_{1\|} = (\nabla\Psi)_{\|}|_{\partial G_2}.$$

Now, since these components of the shielding strengths $\vec{E}_1$ and $\vec{H}1$ on the surface ∂G_2 are known and the total strengths beyond G_2 is identically zero, we find discontinuities of these components of the total strengths:

$$\delta E_{\perp} = \left(E_{1\perp} + (\nabla\Phi)_{\perp}\right)|_{\partial G_2}, \quad \delta\vec{H}_{\|} = \left(\vec{H}_{1\|} + (\nabla\Psi)_{\|}\right)|_{\partial G_2}.$$

Substituting these expressions into the equations (1.2) yields the desired surface densities of electric charge and current on ∂G_2:

(2.3) $$\sigma = \frac{1}{4\pi}\left(E_{1\perp} + (\nabla\Phi)_{\perp}\right)|_{\partial G_2}, \quad \vec{j} = \frac{1}{4\pi}\left(\vec{H}_{1\|} + (\nabla\Psi)_{\|}\right)|_{\partial G_2}.$$

The result obtained provides a certain algorithm of solution of the problem of shielding sources confined in a domain G_1 by charge and current densities on a surface ∂G_2, which consists in the following.

(1) find the strengths $\vec{E}_1$ and $\vec{H}_1$ of the field produced by the original sources

(2) chose a surface ∂G_2 on which the shielding source is to be found

(3) fix normal $E_{1\perp}$, $H_{1\perp}$ and tangential $\vec{E}_{1\parallel}$, $\vec{H}_{1\parallel}$ components of the strengths on the surface

(4) Fourth, find solutions of the Laplace equation (2.1) with boundary conditions (2.2), which disappear at infinity

(5) find normal derivative of Φ and tangential derivative of Ψ on the surface

(6) substitute it into the equation (2.3) and get the desired surface densities of electric charge and current on it

Now we pass to the original problem of the field of an instantly polarized particle in which its dipole moment is shielded causally, without any charge and current densities on the sphere $r = t$ which expands with speed of light. The task is to obtain solutions of the source-free Maxwell equations in which the phenomenon of shielding reveals as a property of electromagnetic field itself. We call this phenomenon the "causal shielding" and presume that it exists and reveals in the field of an instantly polarized particle.

3. The retarded potential in case of instant magnetization

Instant magnetization of a point particle is another possible phenomenon which can be explored with the method of retarded potentials. This phenomenon is interesting for the following reason. Existence of magnetic charge is incompatible with existence of vector potential from which magnetic field can be obtained as in the equation (6.4). In other words, lines of force of magnetic field, which is represented this way, cannot possess any endpoints by construction. Consequently, if a magnetic dipole "switches on" at some moment of time, its magnetic field is non-zero only inside the sphere $r = t$, but lines of force of its magnetic field are closed. In this section we try to clear up how it happens.

Let $\vec{A}(\vec{r})$ is static vector potential of the field of magnetic dipole. If the dipole is "switched on" at the moment of time $t = 0$ one can try the form of vector potential

$$\vec{A}(t, \vec{r}) = \vartheta(t - r)\, \vec{A}(\vec{r})$$

and look how it happens that though the lines of force end up on the sphere $r = t$, they have no endpoints. It turns out that, as expected, in this case the lines of force have no endpoints, simply they suffer a refraction such that each of them consists of three arcs, one starting from the particle and ending up on the surface $r = t$, one lying on the sphere and the closing one starting on the sphere and ending up in the particle. Besides, there appears a δ-function singularity of both strengths:

$$\vec{E} = \delta(t - r)\vec{A}(\vec{r}), \quad \vec{H} = \vec{n} \times \delta(t - r)\, \vec{A}(\vec{r})$$

where $\vec{n}$ stands for the unit vector out from the origin of coordinates (purely radial in spherical coordinates). It is seen that the toroidal field $\vec{E}$ is pure singularity whose time derivative serves as shielding current of displacement on the sphere $t = r$ because it is tangent to the sphere and provides knee points on magnetic lines of force. Note that in this case the strengths satisfy the source-free Maxwell equations. To see this, we have to obtain the entire expressions for the vectors $\vec{A}$, $\vec{E}$ and $\vec{H}$.

There exist two distinct ways the strength of a stationary source-free magnetic field can be represented. Since both curl and divergence of this vector are zero, it can be represented as curl of a vector $\vec{A}$ and as gradient of a scalar Φ. In this case the vector $\vec{A}$ is strictly toroidal, thus possesses only the φ-component in the round cylinder coordinates. Comparison of these two representations yields the following equations for these two potentials,

$$\frac{\partial \Phi}{\partial z} = \frac{1}{\rho}\frac{\partial A}{\partial \rho}, \qquad \frac{\partial \Phi}{\partial \rho} = -\frac{1}{\rho}\frac{\partial A}{\partial z}$$

where A stands for the φ-component of the vector $\vec{A}$, so that $|\vec{A}| = \rho^{-1}A$. As for the scalar potential Φ, it is known, because the duality transformation (6.3) turns the magnetic strength into the electric one, hence, $\Phi = \Phi_{\vec{d}}$ (1.3). Substituting the dipole potential into the equations above, yields partial derivatives of the function $A(z, \rho)$:

$$\frac{\partial A}{\partial z} = -\frac{3z\rho^2}{(z^2 + \rho^2)^{5/2}}, \qquad \frac{\partial A}{\partial \rho} = \frac{\rho(2z^2 - \rho^2)}{(z^2 + \rho^2)^{5/2}}.$$

The solution of these differential equations is

$$(3.1) \qquad A(z, \rho) = \frac{\rho^2}{(z^2 + \rho^2)^{3/2}}.$$

that coincides with the expression for $\rho|\vec{A}|$ obtained in the previous chapter (1.6).

Now, the curl of the vector potential specified by this φ-component, multiplied by the step function $\vartheta\left(t - \sqrt{z^2 + \rho^2}\right)$ is

$$\vec{H} = \nabla\Phi_{\vec{d}} + \vec{n} \times \delta\left(t - \sqrt{z^2 + \rho^2}\right)\vec{A}$$

and the electric strength found from the equation (6.4) is purely singular:

$$\vec{E} = c\delta\left(t - \sqrt{z^2 + \rho^2}\right)\vec{A}$$

and, similarly,

$$\nabla \times \vec{E} = \vec{n} \times \delta'\left(t - \sqrt{z^2 + \rho^2}\right)\vec{A}.$$

Note that time derivative of the magnetic strength is

$$\frac{\partial \vec{H}}{\partial t} = \vec{n} \times \delta'\left(t - \sqrt{z^2 + \rho^2}\right)\vec{A}$$

that coincides with curl of $\vec{E}$. Thus, singular part of the strengths satisfy this Maxwell equation separately. The physical meaning of the sphere expanding with speed of light is not the same as before. Instead of carrying away a dipole moment, this sphere carries a δ-function singularity of electric strength which folds the lines of force of the magnetic field such that they lie on the sphere in accord with Maxwell equations. Since the whole of magnetic flux form the dipole is squeezed onto this surface, the magnetic strength is singular on it.

This result gives a hint how to eliminate endpoints of the lines of force of electric field in case of electric dipole "switched on" at the moment of time $t = 0$. The corresponding electric and magnetic strengths can be obtained now by the duality transformation over the strengths obtained in the case of magnetic polarization. After the transformation we have singular magnetic field of the form $\vec{H} \sim \delta(t - r)$, which plays the same role as electric strength in the case just considered. This will be done in the next section.

4. Elimination of the luminal charge distribution

Now return to "switched on" electric dipole. The task is to find out the solution describing its field without screening by charges and current densities on the expanding sphere $r = t$. This can be done in spherical coordinates as follows. The field to be found is presented partly by $\Phi_{\vec{d}}(r, \theta, \varphi)$ and besides, there must be some vector potential to provide the relevant (toroidal) magnetic strength which contains $\delta(t - r)$ as a factor. Such a vector potential must be non-zero inside the sphere $r = t$, but the corresponding magnetic field must be identically zero there. Consequently, this vector is a pure gradient (6.5) multiplied by $\theta(t - r)$. Indeed, vector potential of this form yields zero magnetic field inside the sphere and does some singular magnetic field on it, which appears when differentiating the step function. Hence, the entire potential has the form

$$(4.1) \qquad \Phi = \theta(t - r)\Phi_{\vec{d}}(r, \theta, \varphi), \quad \vec{A} = \theta(t - r)\nabla\Psi$$

and it remains to find out the function Ψ that is a simple technical problem. Differentiating this potential shows that the function $\Psi(r, \theta, \varphi)$ satisfies the equation

$$(4.2) \qquad \frac{\partial \Psi}{\partial r} = -\Phi.$$

Indeed, the corresponding strengths of electric and magnetic fields are

$$\begin{aligned}
\vec{E} &= \theta(t-r)\nabla\Phi - \delta(t-r)[\Phi\vec{n} + \nabla\Psi] = \\
&= \theta(t-r)\nabla\Phi - \delta(t-r)\nabla_\perp\Psi \\
\vec{H} &= -\delta(t-r)\vec{n} \times \nabla\Psi,
\end{aligned}$$

where $\vec{n}$ is (external) unit normal to the sphere $r = t$, $\nabla_\perp$ stands for components of ∇ orthogonal to it and we have used the equation (4.2). Thus, The field consists of two parts. One of them is exactly dipole field cut on the sphere $r = t$, another is given by strengths of electric and magnetic fields which are non-zero only on the sphere and constitute the δ-function singularity on it. Both its components – electric and magnetic are tangent to the sphere and orthogonal to each other. The electric strength has only θ-component as $\nabla_\perp\Psi$ while the magnetic one has only φ-component as vector product of this and the vector $\vec{n}$.

It remains to check out if these strengths satisfy Maxwell equations. Since two of Maxwell equations are satisfied by construction, the task is to check out that the fields in question are source-free. For this end we evaluate $\nabla \cdot \vec{E}$ and $\nabla \times \vec{H} + \partial\vec{E}/\partial t$. The earlier is

$$\nabla \cdot \vec{E} = \theta(t-r)\Delta\Phi + \delta(t-r)[\vec{n} \cdot \nabla\Phi - \nabla \cdot (\nabla_\perp\Psi)].$$

Note that due to the equation (4.2) the the singular term in this expression is proportional to $\Delta\Phi$. Indeed, due to this equation we have

$$\vec{n} \cdot \nabla\Phi = \frac{\partial\Phi}{\partial r} = \nabla \cdot (\nabla_\perp\Psi) = \nabla_\perp \cdot (\nabla_\perp\Psi)$$

simplifies to the form:

$$\nabla \cdot \vec{E} = \delta(t-r)\left(\frac{\partial\Phi}{\partial r} - \nabla_\perp \cdot \nabla_\perp\Psi\right).$$

To show that this expression is identically zero we use the fact that the operator $\nabla_\perp \cdot \nabla_\perp$ is Laplacian on sphere multiplied by r^{-2}. Multiplying it by r^2 and evaluating its r derivative gives:

$$\nabla \cdot \vec{E} = \ \delta(t-r)\left(\frac{\partial \Phi}{\partial r} - \nabla_\perp \cdot \nabla_\perp \Psi\right) =$$

$$= \ \delta(t-r)\cdot\frac{\partial}{\partial r}\left[r^2\left(\frac{\partial \Phi}{\partial r}\right) - \nabla_\perp \cdot \nabla_\perp \Psi\right] =$$

$$= \ \delta(t-r)\left\{\frac{\partial}{\partial r}\left[r^2\left(\frac{\partial \Phi}{\partial r}\right)\right] + r^2 \nabla_\perp \cdot \nabla_\perp \Phi\right\},$$

hence, the expression in question equals $\triangle \Phi$ thus, is zero. Consequently, $\nabla \cdot \vec{E}$ has zero r-derivative and, due to general considerations, vanishes at infinity, therefore it is safe to claim that it is zero everywhere:

$$\nabla \cdot E = 0.$$

It remains eliminate the shielding current density.

5. Eliminating the luminal current density

Now, consider the closing equation

$$\frac{\partial \vec{E}}{\partial t} + \nabla \times \vec{H} = 4\pi \vec{I}.$$

Substituting the strengths gives:

$$\frac{\partial \vec{E}}{\partial t} = \delta(t-r)\nabla\Phi - \delta'(t-r)\{\nabla_\perp \Psi$$

$$\nabla \times \vec{H} = -\nabla \times [\delta(t-r)\vec{n} \times \nabla\Psi] =$$
$$= \delta(t-r)(\nabla\Phi - \vec{n} \times [\vec{n} \times \nabla\Psi]) + \delta'(t-r)\{\nabla_\perp \Psi\} -$$
$$-\vec{n} \times [\vec{n} \times \nabla\Psi]\}$$

$$\frac{\partial \vec{E}}{\partial t} + \nabla \times \vec{H} = \delta(t-r)(\nabla\Phi - \nabla \times [\vec{n} \times \nabla\Psi]) -$$
$$\delta'(t-r)(\nabla_\perp \Psi - \vec{n} \times [\vec{n} \times \nabla\Psi])$$

where the last term is zero. So, it remains to prove that the expression

$$\nabla\Phi - \nabla \times [\vec{n} \times \nabla\Psi]$$

is zero too. To do so note that

$$\nabla \times [\vec{n} \times \nabla \Psi] \;=\; -\vec{n}(\nabla \cdot \nabla)\Psi + \nabla(\vec{n} \cdot \nabla)\Psi =$$

$$=\; -\vec{n}\Delta\Psi + \nabla\frac{\partial}{\partial r}\Psi = \vec{n}\Delta\Psi - \nabla\Phi.$$

Again, it is seen that differentiating the first term on r gives $\triangle\Phi$ that is zero and, hence, this r-independent term vanishes everywhere, so,

$$\nabla \times [\vec{n} \times \nabla \Psi] = \nabla\Phi,$$

consequently, the δ-term is also zero. Substituting this into the expression for the current density $4\pi\vec{I}$ shows that it is identically zero. However, the form (4.1) does not represent the field of "switched on" dipole, because, though the charge and current densities are eliminated, there also exists the last equation (6.1) with magnetic current as the source which still is not solved, because the strengths do not satisfy it. In any case, the method of retarded strengths cannot be used for obtaining solutions in case of smoothly varying dipole moments and hence, for solving the original H. Hertz problem.

6. Conclusions

We started this chapter with the problem of shielding of arbitrary stationary sources of electric and magnetic field to see what surface densities of electric charge and current on a given surface shield out these sources. After these densities have been found we passed to the problem of causal shielding, supposing that expanding of the surface with speed of light makes it possible to replace the shielding surface densities with a singularities of strengths of the fields. In other words, we tried to replace the shielding current with the current of displacement on the surface, using its expansion.

Now, let us resume the results of this chapter.

(1) Our first modification of the method which replaces retarded potentials with retarded strengths, does not provide the correct solution of the Maxwell equations.

(2) In case of instant magnetization, retarded strengths constitute a quite realistic description of the process of free propagation of lines of force because, at least, endpoints of lines of force field in the case of electric polarization are eliminated.

(3) The speed of propagation equals the speed of light. However, the field has one unusual property: unlike all plane waves the field is apparently non-orthogonal to the direction of propagation.

(4) Another modification of the method of retarded potentials removes the shielding current density by the current of displacement adding a term which acts as a gauge transformation inside the sphere $r = t$, but is identically zero beyond it. The additional term which specifies the gradient transformation, is multiplied by the step function $\vartheta(t)$, therefore, it yields the desired singularities on the expanding sphere.

All this means that no modifications of the method of retarded potential can help to do it, the Maxwell equations are to be solved as they stand, that will be done in the next chapter.

CHAPTER 3

The original Hertz problem and its simplest generalization

1. The problem definition

In this chapter we return to the original Hertz problem and explore possibility to solve by the method of variables separation. As was pointed out above, there are three distinct problems of electromagnetic radiation of a point-like dipole, which, however, are equivalent to each other. Since the problem of electromagnetic field produced under instant polarization of a particle and the original Hertz problem are two of them, these two problems are equivalent and as one of them is solved, to solve another is just a technical problem. However, in our approach, it is not vital if magnitude of the dipole moment draws harmonic oscillations. Any other time dependence is as relevant as this one, therefore, below we consider the simplest generalization of the Hertz problem, assuming that magnitude of the dipole moment is an arbitrary function of time.

So, the generalized Hertz problem considered in this chapter, is defined as follows. A point-like electric or magnetic dipole, suspended at rest, has fixed direction in the space and magnitude of its moment draws harmonic oscillations. The task is to find out electromagnetic field produced by it. This chapter is organized as follows. We start with the case of magnetic dipole because its field can be represented by single component of the vector potential A_φ. First, we find the corresponding strengths and substitute them into the Maxwell equations. This yields the equation for the φ-component of the vector potential, which is similar to the d'Alembert equation. Then we solve this equation, obtain the desired solutions of the Maxwell equations and formulate the general solution of the original Hertz problem.

2. The equation for the field of magnetic dipole

In this section we derive equations for an axially-symmetric electromagnetic field specified by its vector potential $\vec{A}$ in spherical coordinates $\{r, \theta, \varphi\}$. Again, it is easier to do it for the field of magnetic dipole, then find the field of an electric dipole from it by the duality transformation (6.3). Due to the problem under consideration, the vector $\vec{A}$ will be represented by the only non-zero component A_φ, that needs to be defined. The point is that in cylindric coordinates, this component has natural definition, due to which the integral over a circle $C : \rho = \text{const}$ lying in a plane $z = \text{const}$

$$\int_C d\vec{r} \cdot \vec{A} = \int_0^{2\pi} A_\varphi \, d\varphi,$$

so that $|\vec{A}| = \rho^{-1} A_\varphi$. Thus, the field is specified by two functions Φ and A_φ, which depend only on time and two other coordinates, thus, z and ρ in the case of cylindric coordinates. Hereafter we omit the subscript φ and specify an axially-symmetric field as a function $A(z, \rho)$.

The strengths which correspond to the potential $A(z, \rho)$ have the form

$$(2.1) \qquad E_\varphi = \frac{\partial A}{\partial t}, \quad H_r = \frac{1}{r^2 \sin\theta} \frac{\partial A}{\partial \theta}, \quad H_\theta = -\frac{1}{\sin\theta} \frac{\partial A}{\partial r}$$

(see Appendix B). Substituting these strengths into the source-free Maxwell equations (6.1) yields the differential equation the function $A(z, \rho)$:

$$(2.2) \qquad \frac{\partial^2 A}{\partial t^2} - \frac{\partial^2 A}{\partial r^2} - \frac{\sin\theta}{r^2} \frac{\partial}{\partial \theta} \left(\frac{1}{\sin\theta} \frac{\partial A}{\partial \theta} \right) = 0$$

(see the Appendix B, the equation(B.2) for detailed derivation). It leap the eyes that this equation differs from the d'Alembert equation commonly used in this capacity. Evidently enough that solutions of these two equations are different, therefore it is impossible to obtain solutions of the Maxwell equations from solutions of the d'Alembert equation which underlie the method of retarded potentials. The difference will be analyzed after solution of the Hertz problem is constructed.

3. Solution of the Maxwell equations

The equation (2.2) is similar to numerous equations found in standard texts on mathematical physics and possess the important property of separability of variables. The method of variables separation provides particular solutions of this equation, which have the form of products of single variable functions:

$$(3.1) \qquad A(t, r, \theta) = R(r)\Theta(\theta) \cos\omega t.$$

Anticipating things, we assume that the function $\Theta(\theta)$ satisfies the equation

$$\sin\theta \left(\frac{\Theta'(\theta)}{\sin\theta} \right)' + l(l+1)\Theta = 0.$$

To solve it, we use the following trick. We introduce a new function $f(\theta)$ defined as $\Theta' = f \sin\theta$. Substituting this into the equation yields one more relation of the functions f and Θ, so that the equation to be solved turns into the following system:

$$f' = -\frac{l(l+1)}{\sin\theta}\Theta, \quad \Theta' = f \sin\theta.$$

Now, excluding the function Θ leads to the following ordinary differential equation for the function f alone:

$$(3.2) \qquad \frac{1}{\sin\theta}\frac{\mathrm{d}}{\mathrm{d}\theta}\left(\sin\theta\frac{\mathrm{d}f}{\mathrm{d}\theta}\right) + l(l+1)f = 0$$

That is nothing but the Legendre equation [14] for the function $P_l(\cos\theta)$. Consequently,

$$f(\theta) = P_l(\cos\theta), \quad \Theta(\theta) = -\frac{\sin^2\theta}{l(l+1)}P_l'(\cos\theta).$$

Since the function Θ must not have zero under $\theta = \pi/2$, we put $l = 1$.

As the result, we obtain the following ordinary differential equation for the radial function $R(r)$:

$$R'' + \left(\omega^2 - \frac{2}{r^2}\right)R = 0.$$

Maxwell equations have been reduced to this one for the first time by G. Mie in 1908 [11]. Solutions of this equations are Whittaker functions $M_{5/2,\pm 3/2}(\omega r)$ [14]. In case $\omega = 0$ the radial function is just r^{-1} that explains the coincidence mentioned in the Chapt. 1, namely, why the method of Green functions yields the correct result in this case. However, if $\omega \neq 0$, solutions of the Maxwell equations do not contain the retarded time $t - r$ which obligatory appears in the explicit form of retarded potential. This fact explains why the method of retarded potential does not work. Thus, finally, we have the desired solution of the Maxwell equations given by the vector potential (3.1), which has single non-zero component A_φ:

$$A_\varphi = M_{5/2,\pm 3/2}(\omega r)\sin^2\theta\cos\omega t.$$

It remains to represent the strengths of the dipole field, which are solutions of the Maxwell equations. Substituting the vector potential into the equations

(2.1) yields the following expressions of the strengths for magnetic Hertz dipole:

$$(3.3) \qquad
\begin{aligned}
E_\varphi &= -M_{5/2,\pm 3/2}(\omega r)\sin^2\theta\sin\omega t \\
H_r &= \frac{M_{5/2,\pm 3/2}(\omega r)}{r^2}\cos\omega t \\
H_\theta &= 2\omega M'_{5/2,\pm 3/2}(\omega r)\sin\theta\cos\omega t.
\end{aligned}$$

It was absolutely the simplest problem in classical electrodynamics.

4. The field of Hertz dipole in Cartesian coordinates

Dipole fields with their sources placed in one and the same point possess the following property. Any linear combination of such fields has the form of a dipole field. In other words, dipole fields form a vector space as well as the dipole momenta do so. Indeed, since the field equation is linear, the Laplace operator applied to linear combination of dipole fields is to give linear combination of their sources, i. e. dipole moments.

Another explanation of this property consists in the following. The potential Φ produced by a dipole with moment $\vec{d}$ can be obtained from the potential of point-like charge q by applying the vector $\vec{d}$ to the Coulomb potential (1). Consequently, the entire set of dipole fields can be represented as the space formed by all constant vectors that can applied to given Coulomb potential. Hence, to describe the field of dipole whose orientation in the space depends on time it suffices, in general, it suffices to take the sum of fields produced by three non-collinear Hertz dipoles.

In order to introduce a triplet of potentials produced by dipoles whose moments are orthogonal to each other it is natural to represent them in Cartesian coordinates. For this end it is necessary to rewrite down the equations (3.3) this way. The passage from spherical to Cartesian coordinates and the corresponding local frames can be completes as follows: use the identities

$$\begin{aligned}
E_\varphi\,d\varphi &\equiv E_x\,dx + E_y\,dy + E_z\,dz, \\
H_r\,dr + H_\theta\,d\theta &\equiv H_x\,dx + H_y\,dy + H_z\,dz,
\end{aligned}$$

and the Jacobi matrix

$$
\begin{aligned}
d\varphi &= \frac{x\,dy - y\,dx}{x^2 + y^2} \\
d\theta &= \frac{z(x\,dx + y\,dy) - (x^2 + y^2)\,dz}{(x^2 + y^2 + z^2)\sqrt{x^2 + y^2}} \\
dr &= \frac{x\,dx + y\,dy + z\,dz}{\sqrt{x^2 + y^2 + z^2}}.
\end{aligned}
$$

As the result, we obtain Cartesian components of the strengths

$$
\begin{aligned}
E_x &= \frac{M y z^2}{(x^2 + y^2)(x^2 + y^2 + z^2)}\sin\omega t \\
E_y &= -\frac{M z^2 x}{(x^2 + y^2)(x^2 + y^2 + z^2)}\sin\omega t \\
H_x &= \frac{(M + 2\omega M'z)x}{(x^2 + y^2 + z^2)^{3/2}}\cos\omega t \\
H_y &= \frac{(M + 2\omega M'z)y}{(x^2 + y^2 + z^2)^{3/2}}\cos\omega t \\
H_z &= \frac{[M z - 2\omega M'(x^2 + y^2)]}{(x^2 + y^2 + z^2)^{3/2}}\cos\omega t
\end{aligned}
$$
(4.1)

for a unit dipole pointing to z-direction, where M stands for the Whittaker function $M_{5/2,\pm 3/2}(\omega r)$ and M' for its derivative.

5. The field of rotating dipole

The field of a rotating dipole is sum of fields of two Hertz dipoles whose moments lie in orthogonal directions and phases shifted by $\pi/2$. It is easy to compose such a field from the strengths provided in the equations (4.1) for the case of a magnetic dipole rotating about the z-axis:

$$E_x = \frac{Mz^2x}{(x^2+z^2)(x^2+y^2+z^2)}\cos\omega t$$

$$E_y = \frac{Mzx^2}{(y^2+z^2)(x^2+y^2+z^2)}\sin\omega t$$

$$E_z = -\frac{Mx^2y}{(y^2+z^2)(x^2+y^2+z^2)}\sin\omega t -$$

$$-\frac{Mxy^2}{(x^2+z^2)(x^2+y^2+z^2)}\cos\omega t$$

$$H_x = \frac{[Mx - 2\omega M'(y^2+z^2)]}{(x^2+y^2+z^2)^{3/2}}\cos\omega t + \frac{(M+2\omega M'y)x}{(x^2+y^2+z^2)^{3/2}}\sin\omega t$$

$$H_y = \frac{(M+2\omega M'x)y}{(x^2+y^2+z^2)^{3/2}}\cos\omega t + \frac{[My - 2\omega M'(x^2+z^2)]}{(x^2+y^2+z^2)^{3/2}}\sin\omega t$$

$$H_z = \frac{(M+2\omega M'x)z}{(x^2+y^2+z^2)^{3/2}}\cos\omega t + \frac{(M+2\omega M'y)z}{(x^2+y^2+z^2)^{3/2}}\sin\omega t.$$

It is natural to expect that the field possesses an angular momentum about the same axis. Below we try to estimate this quantity.

The density of the z-component of the angular momentum of the field can be obtained by calculating the integral over the space

$$M_z = \int d\vec{r}\,[xP_y - yP_x],$$

where $\vec{P}$ is the Poynting vector of the field. Therefore, to find the angular momentum we need, first of all, to find out the x- and y-components of the Poynting vector. However, we do not need all the expressions of these components because each component of strengths contains oscillating factors which are M, M', $\sin\omega t$ and $\cos\omega t$ whose contributions either are oscillating functions of time or r or make the integrand oscillating such that the integral of the corresponding term either vanishes or is negligibly small. Unlike them, their squares are strictly non-negative and give the main contribution into the integral. Below P_x and P_y stand for contributions of only terms proportional to M^2, $\sin^2\omega t$ and $\cos^2\omega t$. The component $P_x = E_yH_z - E_zH_y$ consists of two terms. Consider the first of them. The first term contains only one expression of this kind, equal to

$$\frac{M^2x^2z}{(y^2+z^2)(x^2+y^2+z^2)^3},$$

but this expression does not contribute the integral because it is odd on the variable z. It remains to consider the product $-E_z H_y$. This product is equal to the sum of the terms proportional to $\sin^2 \omega t$ and $\cos^2 \omega t$, while all the rest are to be ignored. However, the earlier is an entirely even function of all three coordinates, whereas this component of the Poynting vector will be multiplied by y and integrated over the space. Evidently, the integral is zero because the integrand is again an odd function. It remains to consider the $\cos^2 \omega t$ term. This term is entirely odd, so, after multiplying it by y, it remains odd on the variable x, hence, the integral is zero. So is integral of another component of the Poynting vector. Consequently, The field produced by a rotating dipole has zero angular momentum. Otherwise, it would be strange if a point-like object produces a finite angular momentum within a finite period of time. However, as seen, in this case radiation from a rotating dipole cannot explain its frequency decrease.

6. Solutions of the equation for the vector potential in prolate spheroidal coordinates

There exist numerous descriptions of prolate spheroidal coordinates in standard texts on mathematical physics. We start with the following notations of the coordinates: the surfaces $u = $ const are confocal prolate spheroids, the coordinate surfaces $v = $ const are two-sheet round hyperboloids with the same foci placed in the points $z = \pm a$ of the z-axis, hereafter we put $a = 1$, and the metric of the coordinate system $\{u, v, \varphi\}$, where, as usual, φ stands for the angle about the axis. The limiting case of spheroids, $u = 0$ is a straight line segment with the foci as its endpoints, therefore it is convenient to represent the rod as this segment, thus, the rod coincides with the coordinate surface $u = 0$. Below we solve the Maxwell equations in these coordinates.

For this end, we solve the equation for the φ-component of the vector potential, which has the form (for detailed derivation, see the Appendix B, the equation (B.3))

$$(\sinh^2 u + \sin^2 v) \frac{\partial^2 A}{\partial t^2} -$$
$$\left[\sinh u \frac{\partial}{\partial u} \left(\frac{1}{\sinh u} \frac{\partial A}{\partial u} \right) + \sin v \frac{\partial}{\partial v} \left(\frac{1}{\sin v} \frac{\partial A}{\partial v} \right) \right] = 0.$$

Note that the strengths obtained from the φ-component of the vector potential are to be changed by the duality transformation (6.3). As usual, particular solutions of this equation can be obtained by the method of variables separation, which can be applied to the desired function of the form $A(t,u,v) = U(u)V(v)\cos\omega t$. After substituting this product of single variable functions the equation decays into two ordinary differential equations

$$(6.1) \qquad \sinh u \left(\frac{U'}{\sinh u}\right)' + \left(\omega^2 \sinh^2 u - k^2\right) U = 0$$

$$\sin v \left(\frac{V'}{\sin v}\right)' + \left(\omega^2 \sin^2 v + k^2\right) V = 0,$$

where k, where is the separation constant. If $\omega = 0$, the field is stationary, the second equation coincides with the equation (3.2) and its solution is

$$V(v) = -\frac{\sin^2 v}{l(l+1)}\, P_l'(\cos v)$$

and the first equation is similar to it, so, solutions are expressed in terms of Legendre functions of real $P_l(\cos v)$ and imaginary argument $q_l(\cosh u)$, [14]. If $\omega \neq 0$ the solution is unknown, but it is seen that asymptotically it does not coincide with the solution for a Hertz dipole. Indeed, though the coordinate v coincides with the spherical θ, the functions $\Theta(\theta)$ and $V(v)$ are different. It is seen form this fact that shapes of electromagnetic oscillations of a dipole and a rod are different. The ordinary differential equations obtained are apparently related to an interesting class of special functions. However, investigations of these equations goes beyond our goals and hereafter we continue as if these functions are known. Then solutions of the Maxwell equations have the form

$$(6.2) \qquad \begin{aligned} E_u &= U(u)V'(v)\cos\omega t, \\ E_v &= -U'(u)V(v)\cos\omega t, \\ H_\varphi &= -U(u)V(v)\sin\omega t \end{aligned}$$

where $U(u)$ and $V(v)$ are solutions of the equations (6.1). The next task is to connect the strengths with current and charge densities on the rod. Evidently, these densities are non-zero on the rod. The connection between the strengths and densities follows from the Maxwell equations (6.1) and in the case of infinitesimally-thin conductor is given by the equations (6.2).

The axis of symmetry is divided into three parts, two semi-lines with $|z|$ running from 1 to infinity and the rod, on which z runs from -1 to 1. On the rod the coordinate u is zero and v runs from 0 to π and on the semi-lines v is as on the the nearest endpoint of the rod and $u = 0$. Thus, the charge and current densities are non-zero only under $u = 0$ and are functions of t and v.

The current I in a given point of the rod is equal to the limit of the contour integral

$$I = \lim_{u \to 0} \int H_\varphi \, \mathrm{d}\varphi$$

and must be zero in the limit

$$\lim_{v \to 0} \int H_\varphi \, \mathrm{d}\varphi$$

under v equal to 0 or π.

Since right now exact solutions of the equations (6.1) are unknown, it remains to find out approximate analytical solutions for the "wave zone" where $\sinh u$ and $\cosh u$ are essentially larger than one. In this zone we can neglect $\sin^2 v$ in the nominator of the first term of the equation (B.3) and replace $\sinh u$ with the new variable $r = e^u$. As the result, the equation turns into the equation (B.2) solved above, but now we need more general solutions because in this case we do not need to restrict the scope with the single value of the integer l as in the case of Hertz dipole and employ the general solution which is Whittaker function $M_{5/2,\pm(2l+1)/2}(\omega r)$ and which describes electromagnetic field of a conducting rod at large distances.

7. Conclusions

In this section, we resume the results of this chapter.

(1) The classical Hertz problem reduces to a single partial differential equation which is similar to equations found in standard texts on mathematical physics and could be solved the same way.

(2) Solution of his equation is found and the explicit form of the field strengths is presented.

(3) This solution resembles solutions of all known solutions of the equations of mathematical physics, in particular, it has the form of product of single variable functions which depend on time and spatial coordinates. One of these factors, which depend on the spherical radius, is the Whittaker function $M_{5/2,\pm3/2}(\omega r)$ which cannot appear under constructing the solution by the method of retarded potentials for a Hertz dipole.

(4) A thin conducting rod which carries oscillating charge and current densities, represents the simplest generalization of the notion of Hertz dipole. The problem of electromagnetic field produced by these oscillations, reduces to the Maxwell equations, which, in turn, reduce to a single partial differential equation similar to that appeared in the classical Hertz problem.

(5) The equation obtained possesses the same properties, but its boundary conditions are specified in prolate spheroidal coordinates. Particular solutions of this equation form an orthonormal basis in the Hilbert space of electromagnetic fields and specify an orthonormal basis in the functional space of charge and current densities on the rod.

(6) Solutions of this generalization of the classical Hertz problem would describe electromagnetic fields of the dipole antenna, proviso that a theory of special functions defined by the equations (6.1) is created.

(7) Asymptotical solutions for the field of conducting rod, are represented as the φ-component of the vector potential which yields the dual to the strengths, contain the Whittaker functions $M_{5/2,\pm(2l+1)/2}(\omega r)$:

$$A_\varphi = M_{5/2,\pm(2l+1)/2}(\omega r) \frac{\sin^2\theta}{l(l+1)} P_l'(\cos\theta)\cos\omega t$$

and describe the field at large enough distances from the rod.

(8) These solutions are not "pure states" of the field and constitute some superpositions of the solutions considered above (6.2). The solution exposes l conical directions from the source, in which amplitude of the wave is zero.

CHAPTER 4

Magnetic burst

1. The problem definition

In this chapter we return to the problem of free propagation of lines of force, but this time the lines have no endpoints because they are closed by construction. The phenomenon we are considering below is a kind of magnetic burst which can theoretically occur under the following circumstances. An ideal toroidal coil is known to produce magnetic fields which are completely confined inside, so that the field in the outer space is identically zero. Suppose, such a coil is made of a highly cooled superconductor and connected to a source of direct current, thus, is actually pumped with stationary magnetic field. If, at a moment of time $t = 0$ it loses its conductivity, say, due to fast heating, the magnetic field shoots up so that its closed lines of force propagate freely in the space. The task is to describe the electromagnetic field to be observed under $t > 0$. This problem is too complicated to be solved analytically, therefore we restrict it with solving only the Maxwell equations for the magnetic field. The particular solutions can be used afterwards for completing this task because they constitute the complete functional basis of toroidal waves.

The property of toroidal coils to produce magnetic fields only inside follows from the fact that the only non-trivial solution, of the equation

$$\nabla \times \vec{H} = 0$$

which vanishes at infinity for strictly toroidal magnetic strengths is

$$(1.1) \qquad\qquad\qquad H_\varphi = C.$$

This fact signifies also that this property of closed axially-symmetric coils is invariant with respect to all possible deformations which do not break the axial symmetry. In other words, it is not obligatory that the coil is exactly of toroidal shape, its meridian section can be of arbitrary form. The most convenient form from mathematical point of view, is an oblate spheroid. A spheroidal coil can be made as follows: there is a strong direct current along its axis and the circuit is closed by the spheroidal surface. Then, if the current from the south to north pole along the axis, on the surfaces it is from the north to south pole. Due to symmetry considerations, its φ-component is zero, hence, the strength of the magnetic field produced by the currents, is as in the equation (1.1) inside the spheroid and zero beyond.

Hence, the task is to describe free propagation of this field in the outer space, where the constant C is to be replaced by some function of three coordinates including time. Though in this case we need only a particular solution and the method of Green functions would be quite satisfactory, as was shown above, results obtained by this method are not reliable, therefore we need solutions of the Maxwell equations obtained another way.

2. Solutions of the Maxwell equations for toroidal magnetic fields

Solutions of the Maxwell equations for toroidal magnetic fields describe the phenomenon of free propagation of circular lines of force. To obtain them it is convenient to solve the equation for the φ-component of the vector potential that yields poloidal magnetic and toroidal electric strengths, then apply the duality transformation. To meet the boundary conditions on a spheroid, we chose spheroidal coordinates, in which the outer boundary of the coil is given as one of coordinate spheroids. Equation for the φ-component of the vector potential represented as $A = U(u)V(v)\cos\omega t$ reduces to ordinary differential equations

$$\cosh u \left(\frac{U'}{\cosh u} \right)' + (\omega^2 \cosh^2 u - k^2)U = 0$$

$$\cos v \left(\frac{V'}{\cos v} \right)' + (k^2 - \omega^2 \cos^2 v)V = 0,$$

that is similar to the equations (6.1), so their solutions are unknown. However, the complete solutions can be obtained in the particular case $k = 0$:

$$U(u) = c_1 \cos(\omega \sinh u) + c_2 \sin(\omega \sinh u)$$

$$V(v) = c_3 \cosh(\omega \sin v) + c_4 \sinh(\omega \sin v).$$

Due to the definition of the φ-component of the strengths we use, the component H_φ vanishes on the axis $v = \pi/2$, that implies $c_4 = 0$, however, at the same time, this function must be even on $\sin v$. The solutions obtained do not meet both of these two conditions, consequently, the case $k = 0$ does not provide any descriptions of the magnetic burst. In fact, the solutions obtained describe various forms of the phenomenon of free propagation of toroidal lines of force, however, the solutions obtained do not constitute a complete basis.

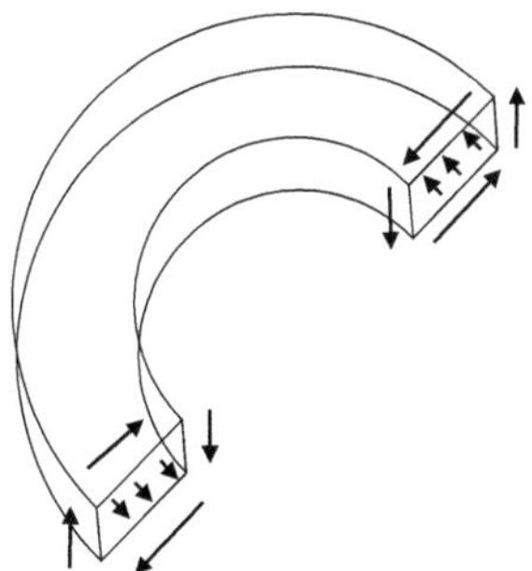

FIGURE 1. A scheme of a hub coil. Arrows specify directions of current and magnetic strength.

3. Solutions for the TH-waves

The idea of squeezed coil is another idealization of that of toroidal coil. A squeezed coil is a conducting shell which is ring plate in form with toroidal magnetic field whose strength is as in the equation (1.1) confined inside an infinitesimally-thin ring plate and identically zero beyond it. The task is, as before, to describe free propagation of the lines of force of magnetic field in case of complete conductivity loss by the shell at some moment of time $t = 0$. We call this phenomenon "burst of a squeezed coil".

To obtain the relevant solutions of the Maxwell equations we solve the equation (B.1) for toroidal vector potential which yields toroidal electric and poloidal magnetic strengths, then apply the duality transformation 6.3). Partial solutions of this equation have the form

$$A_\varphi = l\rho J_1(l\rho)\cos\omega t\cos kz, \quad \omega^2 - k^2 - l^2 = 0$$

and, after the duality transformation, specify the φ-component of the magnetic strength

$$(3.1) \qquad\qquad H_\varphi^{\omega,l} = \rho J_1(l\rho)\cos\omega t\cos kz,$$

where some constant coefficients, particularly, ω, are omitted. The total strength is a linear combination of these particular solutions with coefficients to be found from the equation 1.1).

Let the ring plate has internal and external radii equal to a and b correspondingly and the plate lies on the plane $z = 0$. Then, at the moment of time $t = 0$ the magnetic field is non-zero only on the plane and its φ-component has the form

$$(3.2) \qquad H_\varphi = \begin{cases} H_0, & \rho \in [a, b] \\ \\ 0, & \rho \notin [a, b]. \end{cases}$$

Under $t > 0$ depends on three coordinates as a linear combination of particular fields (3.1) which is be recovered from its initial form

$$\rho^{-1} H_0 = - \int_0^\infty dl\, l\, g(l)\, J_1(l\rho)$$

where we put $t = z = 0$. Suppose that the Fourier-Bessel transform of this component $g(l)$ is represented as the following Fourier integral:

$$(3.3) \qquad g(l) = \frac{2\pi}{l} \int_0^\infty du\, \phi(u) \cos lu$$

where we put $\phi = 0$ under $u \geq b$. Then, substituting this expansion into the expansion of the function $\rho^{-1} H_0$ and integrating over the variable l allows us to rewrite it in the form of integral representation

$$(3.4) \qquad \frac{H_0}{2\pi\rho} = \int_\rho^b \frac{du\, u\, \phi(u)}{\sqrt{u^2 - \rho^2}}.$$

The next step is to solve this integral equation and obtain the explicit form of the function $\phi(u)$.

For this end, we multiply both sides by $\rho/\sqrt{\rho^2 - s^2}$ and integrate over the variable ρ from s to b. The double integral in the right-hand side transforms as follows:

$$\int_s^b \frac{d\rho\, \rho}{\sqrt{\rho^2 - s^2}} \int_\rho^b \frac{du\, u\, \phi(u)}{\sqrt{u^2 - \rho^2}} = \int_s^b du\, u\, \phi(u) \int_s^b \frac{d\rho\, \rho}{\sqrt{(u^2 - \rho^2)(\rho^2 - s^2)}} =$$

$$= \frac{1}{2} \int_s^b du\, u\, \phi(u) \int_0^1 \frac{dv}{\sqrt{v(1 - v)}} = \frac{\pi}{2} \int_s^b du\, u\, \phi(u).$$

Transformation of the left-hand side is simpler: if $s \geq a$,

$$\int_s^b \frac{\mathrm{d}\rho\rho}{\sqrt{\rho^2 - s^2}} \frac{H_0}{2\pi\rho} = \frac{H_0}{2\pi} \operatorname{arccosh}\frac{b}{s}$$

otherwise,

$$\int_s^b \frac{\mathrm{d}\rho\rho}{\sqrt{\rho^2 - s^2}} \frac{H_0}{2\pi\rho} = \int_a^b \frac{\mathrm{d}\rho\rho}{\sqrt{\rho^2 - s^2}} \frac{H_0}{2\pi\rho} = \frac{H_0}{2\pi} \left(\operatorname{arccosh}\frac{b}{s} - \operatorname{arccosh}\frac{a}{s} \right).$$

As the result, we obtain the following equations:

$$\frac{\pi}{2} \int_s^b \mathrm{d}uu\phi(u) = \begin{cases} \frac{H_0}{2\pi} \left(\operatorname{arccosh}\frac{b}{s} - \operatorname{arccosh}\frac{a}{s} \right), & s < a \\[2ex] \operatorname{arccosh}\frac{b}{s}, & s \in [a, b]. \end{cases}$$

Finally, the function $\phi(s)$ comes out in an non-analytical form

$$(3.5) \qquad \phi(s) = \begin{cases} \dfrac{H_0}{\pi^2 s} \left(\dfrac{1}{\sqrt{b^2 - s^2}} - \dfrac{1}{\sqrt{a^2 - s^2}} \right), & s < a \\[3ex] \dfrac{H_0}{\pi^2 s \sqrt{b^2 - s^2}}, & s \in [a, b]. \end{cases}$$

Now, substituting the functions ϕ into the equation (3.3) yields the function $g(l)$ which provides an l-spectrum, but tells nothing about wavelengths. Consequently, highly squeezed coil is a too idealized object which cannot be used as a model of a burster. To get a more realistic model we need a more realistic model, therefore we pass to model of a hub coil shown on the Fig. 1.

4. Burst of a hub coil

In case of burst of a hub coil the equations (3.1-3.5) remain in force, but initial conditions are different. Unlike a squeezed toroidal coil, a hub has a finite height. Let its top and bottom lie in planes $z = \pm h$. Then under $t = 0$ the magnetic strength has the form

$$\rho^{-1} H_0 = -\vartheta(h - |z|) \int_0^\infty \mathrm{d}llg(l)J_1(l\rho)$$

where the box function $\vartheta(h - |z|)$ is to be expanded over the cosines of z. The Fourier-transform of this function is known [**15**]:

$$\vartheta(h - |z|) = 2 \int_0^\infty \frac{\sin kh \cos kz}{k},$$

hence, at $t = 0$ the strength is

$$\rho^{-1} H_0 = -2 \int_0^\infty \frac{\sin kh \cos kz}{k} \int_0^\infty dll\, g(l) J_1(l\rho)$$

where the function $g(l)$ is given by the equation (3.3), the function $\phi(u)$ by the equation (3.5). Finally, we have the following amplitude as a function of frequency of radiation of a magnetic burst of a hub coil:

$$f(\omega) = \frac{\sin h\sqrt{\omega^2 - l^2}}{\sqrt{\omega^2 - l^2}} g(l).$$

Other observables like energy spectra if given frequency bands can be obtained from this function. Since this function has zeros, spectra of magnetic bursts seem to possess the same peculiarity. This peculiarity can be used for testing bursters models in astronomy.

5. Conclusions

Results presented in this chapter, consist in the following.

(1) The phenomenon of free propagation of lines of force, whose shape is fixed, is interesting by itself. In this chapter we have considered the phenomenon of burst of toroidal magnetic field.

(2) Maxwell equations for toroidal fields separate in round cylinder, spherical and spheroidal coordinates and reduce to ordinary differential equations.

(3) Maxwell equations for toroidal fields separate in round cylinder, spherical and spheroidal coordinates and reduce to ordinary differential equations.

(4) The burst of a hub coil can be described in round cylinder coordinates, but this description contains very complicated integrals. Calculation of these integrals also is a subject of special investigation, which goes beyond the scope of our book.

(5) Nevertheless, a model of this phenomenon is built and amplitude as a function of frequency is obtained in terms of the same integrals.

(6) As a result, we have a qualitative description of the spectra which exposes the following peculiarity of frequency spectra: most likely, they have zeros. This peculiarity can be used for estimating sizes of bursters or for testing their models.

CHAPTER 5

Conclusions

Like the notion of point-like electric charge, which is the most fundamental source of electric filed in electrostatics, the notion of point-like dipole with varying dipole moment plays the role of the most fundamental source of electromagnetic field in classical electrodynamics. As a source of non-stationary electromagnetic fields, a point-like dipole appears in, at least three fundamental forms, a Hertz dipole, a rotating dipole and an instantly polarized particle. Though the problems of field produced by these three sources are equivalent, their solutions need different approaches, particularly, the latter seem to be the simplest illustration of the meaning of retarded potential. However, geometric appearance of its field exposes an unexpected property. It turns out that lines of force of its field have endpoints which cannot exist unless there are electric or magnetic charges in the space. On the other hand, due to the problem definition, there are neither electric nor, the rather, magnetic charges in the space, so, the problem of an instantly polarized particle required more thorough studies.

The firs idea is that the method of retarded potentials needs some minor modification to remove the charge density which appears on the luminal sphere $r = t$. It seems to be clear how to modify it, because, on one hand, the same should happen in the case of magnetic dipole, and on the other hand, if magnetic field is represented in terms of vector potential, its lines of force cannot have endpoints by construction. Indeed, representation of the strengths via the vector potential and replacing retarded potentials with retarded strengths removes endpoints of the lines of force and the problem seems to be solved. However, it turns out that this modification of the method of retarded potential does not provide solutions of the Maxwell equations in the case of smoothly varying dipole moment. Nor another possible modification which contains a gradient transformation cut with the step function $\vartheta(t - r)$ helps and as our ideas of possible modifications of the method of retarded potential are exhausted, it remains to conclude that Maxwell equations should be solved as they stand. After all, Maxwell equations are as linear as numerous equations of mathematical physics which are always being solved by the method of variables separation, whose solutions constitute an orthonormal basis in a Hilbert space and provide the most general solutions of the corresponding problems. Only Maxwell equations are excepted from this general scheme. Maxwell equations have not been solved this way even in the simplest case when the field is produced by a point-like dipole. This situation apparently needed to be improved.

Usually, the main difficulty encountered when solving the Maxwell equations is to decouple the components. But in the case of point-like dipole the field potential has only one non-zero component A_φ, so this problem does not appear. The equation for this component could well be derived in the framework of more or less standard mathematics. The only difficulty arises because one needs to find curl of a vector whose components are referred to non-parallel frames which naturally appear in spherical coordinates. It has been shown in the literature (see, for example, the books [**12, 13**]) how to do it in terms of the standard vector analysis. In this book we have exposed derivation of the equation in terms of exterior differential forms that gives exactly the same result. This exposition is more useful because it can be applied not only in round cylinder or spherical coordinates, but also in all possible coordinate systems for the space or the space-time. The first thing leaping the eyes is that the genuine equation for the φ-component completely differs from the d'Alembert equation used and the underlying base of the theory of retarded potentials.

Moreover, the equation obtained separates and solves in terms of special functions. The exact solution for the field of Hertz dipole is expressed in Whittaker functions and has nothing to do with retarded potentials, though coincide with them at infinity. As solution of the H. Hertz problem was obtained, we passed to its generalizations. Usually a point-like dipole serves as a model of a simple antenna which is a thin conducting rod in form. So, the nearest generalization of this problem is that of electromagnetic oscillations of a thin conducting rod. Solution of this problem requires usage of prolate spheroidal coordinates in which again one needs to take curl of a vector referred to the frames which appear in this coordinate system. This time the standard vector analysis cannot help and the only way to do it, is to apply the theory of exterior differential forms.

The corresponding vector potential is represented by the only non-zero component A_φ and we derived the corresponding equation by replacing it with the 1-form $\alpha = A\,\mathrm{d}\varphi$. The equation separates and reduces to uncoupled ordinary differential equations whose solutions are unknown transcendental functions which hardly could appear in the form of retarded potentials. However, the solution is outlined and can be used for the need of the antennas theory. In this case the theory took the standard form of an ordinary linear theory, in which particular solutions of the main equation constitute a functional basis of "pure states" in a Hilbert space as in quantum mechanics.

Besides, there exists one more phenomenon of free propagation of lines of force, but unlike that of instantly polarized particle, these lines have no endpoints because they are closed by definition. Moreover, unlike the field of instantly polarized particle, this field is nowhere stationary. Free propagation of round lines of force occurs after magnetic burst. An initially pumped closed coil can release its magnetic field if it suddenly loses conductivity that may well be realized practically. Its field remains toroidal and propagates in the outer space freely. To describe it one needs the corresponding solutions of the Maxwell equations. Solutions of these ordinary differential equations are known only for round cylinder and spherical coordinates. As for spheroidal ones, a special investigation is needed to create a theory of the class of corresponding special functions.

Besides, there exists an opportunity to describe a burst of a hub coil, because in this case the field equation can be this solved in round cylinder coordinates and its particular solutions are products of well-known functions. However, this description contains very complicated integrals. Calculation of these integrals also is a subject of special investigation, therefore the only goal which we managed to reach towards composing a theory of magnetic burst, is finding the following peculiarity of their spectra. It turns out that spectra of magnetic burst have zeros which, theoretically, make it possible to estimate size of a burster. So, in this book we have exposed the whole of theory of the Hertz dipole and related phenomena.

Appendices

1. Appendix A. Calculation of $\nabla \cdot \vec{E}$ for retarded potential of Hertz dipole

Divergence of the strength $\vec{E}$ in this case is ([5], equation (6.32))

$$(.1) \qquad \nabla \cdot \vec{E} = \triangle \Phi + \frac{\partial^2 A_z}{\partial t \partial z},$$

where Φ and A_z are given by the equations (2.2). These components of the potential yield totally four distinct expressions with powers of ω from 0 to 3 as a common factor. They will be considered separately. First, we consider the terms which are identically zero. One of them corresponds to the power 0 and another has the common factor ω^3. The earlier is identically zero

$$\cos \omega(t - r) \, \triangle \left(-\frac{z}{r^3} \right) \equiv 0$$

because the Laplacian is taken of the static potential of a dipole. Now we need the following auxiliary calculations:

$$(.2) \qquad \triangle \sin \omega(t - r) = -\frac{2\omega \cos \omega(t - r)}{r} - \omega^2 \sin \omega(t - r).$$

$$\triangle \cos \omega(t - r) = \frac{2\omega \sin \omega(t - r)}{r} - \omega^2 \cos \omega(t - r).$$

Thus, right now we need only the second term in the first line. Similarly,

$$(.3) \qquad \frac{\partial^2}{\partial t \partial z} \sin \omega(t - r) = \omega^2 \frac{\partial r}{\partial z} \sin \omega(t - r)$$

Substituting these results and the components of the potential (2.2) into the equation (.1), yields for the ω^3-term:

$$\omega^3 \left(-\frac{z \sin \omega(t - r)}{r^2} + \frac{z \sin \omega(t - r)}{r^2} \right) \equiv 0$$

and it remains to calculate the ω- and ω^2-terms.

65

The ω^2-term consists of contributions of Φ, which is

$$2\omega \left(\nabla \frac{z}{r^2} \cdot \nabla \sin \omega(t-r) \right) + \frac{z}{r^3} \omega^2 \cos \omega(t-r)$$

where the second term comes from one of equations (.2), and contribution of A, which is

$$\omega^2 \sin \omega(t-r) \frac{\partial}{\partial z} \frac{1}{r}.$$

Calculation of the earlier yields the following:

$$\frac{2\omega}{r^2} \nabla z \cdot \nabla \sin \omega(t-r) - \frac{4\omega z}{r^3} \nabla r \cdot \nabla \sin \omega(t-r) +$$

$$+ \frac{\omega^2 z}{r^3} \omega^2 \cos \omega(t-r) = -\frac{3z}{r^3} \omega^2 \sin \omega(t-r).$$

Contribution of the spatial component differs from it only by a numerical coefficient and as the result the ω^2-term is

$$\frac{2\omega^2 z}{r^3} \omega^2 \sin \omega(t-r) \neq 0.$$

The ω-term is non-zero too. Note that the component A_z of the potential does not contribute it, so, the desired expression is

$$\omega z \sin \omega(t-r) \, \triangle \, \frac{z}{r^2} - 2\nabla \left(\frac{z}{r^3} \cdot \nabla \cos \omega(t-r) \right)$$

that is

$$-\frac{2\omega z}{r^4} \sin \omega(t-r) \neq 0.$$

Finally, we have divergence of the strength $\vec{E}$:

$$(.4) \qquad \nabla \vec{E} = \frac{2\omega^2 z}{r^3} \omega^2 \sin \omega(t-r) - \frac{2\omega z}{r^4} \sin \omega(t-r).$$

This expression specifies electric charge density needed to produce electric strength corresponding to the retarded potential of a Hertz dipole (2.2). In almost all books on classical electrodynamics the retarded potential is represented in erroneous form which does not contain the time component Φ as it should be in case of pure radiation. However, in this form the divergence of $\vec{E}$ is non-zero too. This value can be calculated straightforwardly with use of the formula for divergence of a vector in spherical coordinates found in the book [12], equation (92) on the page 52. Divergence of this component has the form

$$\frac{1}{r \sin \theta} \frac{\partial}{\partial \theta} (E \sin \theta)$$

and since other components are zero, this expression is exactly $\nabla \cdot \vec{E}$. It can be zero only if E_θ is proportional to inverse $\sin\theta$. However, the expression for $\vec{E}$ contains only positive powers of $\sin\theta$. Consequently, $\nabla \cdot \vec{E} \neq 0$.

2. Appendix B. Reduction of the equation for A_φ in various coordinate systems

When working with a non-scalar fields in a curvilinear coordinates in the framework of standard vector analysis, the problem arises to apply the standard vector operator ∇ properly. This problem never arises in the framework of the analysis of exterior differential forms. Therefore, we demonstrate reduction of the Maxwell equation. for the field specified by its only non-zero component A_φ of the vector potential. The result can easily be translated to the language of the standard vector analysis.

First, we replace the vector potential with the corresponding 1-form α:

$$\alpha = A_\varphi \, d\varphi$$

and take its exterior derivative to obtain the 2-form of strengths:

$$d\alpha = \frac{\partial A}{\partial t} \, dt \wedge d\varphi - \frac{\partial A}{\partial z} \, d\varphi \wedge dz + \frac{\partial A}{\partial \rho} \, d\rho \, d\varphi.$$

The second step is to apply the asterisk conjugation which is equivalent to the transformation of duality (6.3):

$$^* d\alpha = \frac{1}{\rho} \frac{\partial A}{\partial \rho} \, dt \wedge dz - \frac{1}{\rho} \frac{\partial A}{\partial z} \, dt \wedge d\rho + \frac{1}{\rho} \frac{\partial A}{\partial t} \, dz \wedge d\rho.$$

Now, taking the exterior derivative of the result yields the left-hand side of two Maxwell equations whereas two others are completed by construction:

$$d\,^* d\alpha = \left[\frac{\partial}{\partial \rho} \left(\frac{1}{\rho} \frac{\partial A}{\partial \rho} \right) + \frac{1}{\rho} \frac{\partial^2 A}{\partial z^2} - \frac{1}{\rho} \frac{\partial^2 A}{\partial t^2} \right] dt \wedge dz \wedge d\rho.$$

This 3-form is nothing but the current density which is assumed to be zero. Consequently, the function $A(t, z, \rho)$ which is equal to the φ-component of the vector potential of pure radiation, satisfies the equation

$$(B.1) \qquad \frac{\partial^2 A}{\partial t^2} - \frac{\partial^2 A}{\partial z^2} - \rho \frac{\partial}{\partial \rho} \left(\frac{1}{\rho} \frac{\partial A}{\partial \rho} \right) = 0.$$

The same reduction can be completed in other coordinate systems. Below we do it in spherical coordinates $\{r, \theta, \varphi\}$. We start with the same 1-form α and take its exterior derivative which is

$$\mathrm{d}\alpha = \frac{\partial A}{\partial t}\,\mathrm{d}t \wedge \mathrm{d}\varphi - \frac{\partial A}{\partial r}\,\mathrm{d}\varphi \wedge \mathrm{d}r + \frac{\partial A}{\partial \theta}\,\mathrm{d}\theta \wedge \mathrm{d}\varphi$$

the 2-form of strengths. Now, asterisk conjugation of the result yields the dual strengths:

$$^{*}\mathrm{d}\alpha = \frac{1}{c\sin\theta}\frac{\partial A}{\partial t}\,\mathrm{d}r \wedge \mathrm{d}\theta - \frac{1}{\sin\theta}\frac{\partial A}{\partial r}\,\mathrm{d}t \wedge \mathrm{d}\theta + \frac{1}{r^2\sin\theta}\frac{\partial A}{\partial \theta}\,\mathrm{d}t \wedge \mathrm{d}r.$$

Due to the Maxwell equations, the 3-form of current is the exterior derivative of this 2-form

$$\mathrm{d}\,^{*}\mathrm{d}\alpha = \left[-\frac{1}{c\sin\theta}\frac{\partial^2 A}{\partial t^2} + \frac{1}{\sin\theta}\frac{\partial^2 A}{\partial r^2} + \frac{1}{r^2}\frac{\partial}{\partial\theta}\left(\frac{1}{\sin\theta}\frac{\partial A}{\partial\theta}\right) \right] \mathrm{d}t \wedge \mathrm{d}r \wedge \mathrm{d}\theta,$$

and, as this 3-form is zero, we obtain the following differential equation for the function $A(t, r, \theta)$:

$$(\text{B.2}) \qquad \frac{\partial^2 A}{\partial t^2} - \frac{\partial^2 A}{\partial r^2} - \frac{\sin\theta}{r^2}\frac{\partial}{\partial\theta}\left(\frac{1}{\sin\theta}\frac{\partial A}{\partial\theta}\right) = 0.$$

Now, we repeat this derivation in prolate spheroidal coordinates $\{u, v, \varphi\}$, starting with the same 1-form α. Its exterior derivative, which represents the strengths is

$$\mathrm{d}\alpha = \frac{\partial A}{\partial t}\,\mathrm{d}t \wedge \mathrm{d}\varphi - \frac{\partial A}{\partial u}\,\mathrm{d}\varphi \wedge \mathrm{d}u + \frac{\partial A}{\partial v}\,\mathrm{d}v\,\mathrm{d}\varphi.$$

The asterisk conjugation which must be completed with account of the explicit form of the prolate spheroidal metric

$$\mathrm{d}s^2 = (\sinh^2 u + \sin^2 v)(\mathrm{d}u^2 + \mathrm{d}v^2) + \sinh^2 u \sin^2 v\,\mathrm{d}\varphi^2,$$

yields this 2-form $^{*}\mathrm{d}\alpha$ of dual strengths:

$$^{*}\mathrm{d}\alpha = -\frac{\sinh^2 u + \sin^2 v}{\sinh u \sin v}\frac{\partial A}{\partial t}\,\mathrm{d}u \wedge \mathrm{d}v -$$
$$\frac{1}{\sinh u \sin v}\frac{\partial A}{\partial u}\,\mathrm{d}t \wedge \mathrm{d}v + \frac{1}{\sinh u \sin v}\frac{\partial A}{\partial v}\,\mathrm{d}t \wedge \mathrm{d}u,$$

and exterior derivative of the latter equals the 3-form of current. We calculate this 3-form and equate it to zero:

$$
\mathrm{d}^* \mathrm{d}\alpha = \left\{ -\frac{\sinh^2 u + \sin^2 v}{\sinh u \sin v}\frac{\partial^2 A}{\partial t^2} + \frac{1}{\sinh u \sin v} \times \right.
$$

$$
\left. \times \left[\frac{1}{\sin v}\frac{\partial}{\partial u}\left(\frac{1}{\sinh u}\frac{\partial A}{\partial u}\right) + \frac{1}{\sinh u}\frac{\partial}{\partial v}\left(\frac{1}{\sin v}\frac{\partial A}{\partial v}\right) \right] \right\} \times
$$

$$
\times \, \mathrm{d}t \wedge \mathrm{d}u \wedge \mathrm{d}v = 0
$$

and obtain the equation for the function $A(t, u, v)$:

$$
\text{(B.3)} \qquad (\sinh^2 u + \sin^2 v)\frac{\partial^2 A}{\partial t^2} -
$$

$$
\left[\sinh u\frac{\partial}{\partial u}\left(\frac{1}{\sinh u}\frac{\partial A}{\partial u}\right) + \sin v\frac{\partial}{\partial v}\left(\frac{1}{\sin v}\frac{\partial A}{\partial v}\right) \right] = 0.
$$

Reduction of the equation for toroidal strengths in oblate spheroidal coordinates is similar.

Bibliography

[1] Deutsch A. J. The electromagnetic field of an idealized star in rigid rotation in vacuo.// Ann. Astrophys.– 1955.– V.1.– p.1-10.

[2] Goldreich P., Julian W.H. Pulsar electrodynamics.// Astrophys. J.– 1969.– V.157, No.2.– p.869-880.

[3] Michel, F.C. and Li, H.// Electrodynamics of neutron stars. // Phys. Rep.– 1999.– V.318, pp. 227-297

[4] Feynman R.P., Leighton R.B., Sands M. *The Feynman Lectures on Physics* Reading, Mass., 1964

[5] Jackson, J.D., *Classical electrodynamics.* (New York, NY, Wiley, 1962.) pp. 269-271

[6] Marion J.B. *Classical Electromagnetic Radiation.* (Academic Press, N.Y. 1965) pp. 221, 235.

[7] Griffiths D.J. *Introduction to electrodynamics.* Prentice-Hall of India, 2004, pp. 444-451

[8] Born M., Wolf E. *Prnciples of Optics* Cambridge University Press, Cambridge, 2005, p. 76

[9] Sommerfeld A. *Elektrodynamik* Verlag Harri Deutsch, 1988, p. 290

[10] de Witt B. *The Global Approach to Quantum Field Theory. Vol. 1* Clarendon Press, Oxford, 2003, p. 330

[11] Mie G. Beiträge zür optik trüber medien, speziell kolloidaler metallösungen.// Annalen der Physik – 1908. – V., No 23 – p. 377-445.

[12] Stratton J.A. Electromagnetic Theory. IEEE Press, 2007, p. 52

[13] Schey H.M. div, grad, curl *and all that. An informal text on vector calculus.* W.W. Norton& Company,1997

[14] Abramowitz M., Stegun I., eds., *Handbook of Mathematical Functions with Formulas, Graphs, and Mathematical Tables*, New York: Dover Publications, 1972

[15] http://www.thefouriertransform.com

yes I want morebooks!

Buy your books fast and straightforward online - at one of the world's fastest growing online book stores! Environmentally sound due to Print-on-Demand technologies.

Buy your books online at

www.get-morebooks.com

Kaufen Sie Ihre Bücher schnell und unkompliziert online – auf einer der am schnellsten wachsenden Buchhandelsplattformen weltweit!
Dank Print-On-Demand umwelt- und ressourcenschonend produziert.

Bücher schneller online kaufen

www.morebooks.de

OmniScriptum Marketing DEU GmbH
Bahnhofstr. 28
D - 66111 Saarbrücken
Telefax: +49 681 93 81 567-9

info@omniscriptum.com
www.omniscriptum.com

Printed by Books on Demand GmbH, Norderstedt / Germany